RURAL CANADA

RURAL CANADA

Rural Sociological Perspectives

Satadal Dasgupta

Canadian Studies
Volume 3

The Edwin Mellen Press
Lewiston•Queenston•Lampeter

Library of Congress Cataloging-in-Publication Data

Dasgupta, Satadal, 1937-
Rural Canada : rural sociological perspectives / Satadal Dasgupta.
p. cm. -- (Canadian studies ; v. 3)
Includes bibliographical references and index.
ISBN 0-7734-7431-5
1. Canada--Rural conditions. 2. Agriculture--Canada. 3. Rural-urban migration--Canada. I. Title. II. Canadian studies (Lewiston, N.Y.) ; v. 3.

HN103.5 .D37 2001
307.72'0971--dc21

2001030017

This is volume 3 in the continuing series
Canadian Studies
Volume 3 ISBN 0-7734-7431-5
CS Series ISBN 0-88946-197-X

A CIP catalog record for this book is available from the British Library.

Revised Edition

The Edwin Mellen Press
Box 450
Lewiston, New York
USA 14092-0450

The Edwin Mellen Press
Box 67
Queenston, Ontario
CANADA L0S 1L0

The Edwin Mellen Press, Ltd.
Lampeter, Ceredigion, Wales
UNITED KINGDOM SA48 8LT

Printed in the United States of America

For Krishna, Kaberi, Mohua, and Keya
and Juan and Adario

Other Books by the Author

The Community in Canada: Rural and Urban

Diffusion of Agricultural Innovations in Village India

Caste, Kinship and Community: Social System of a Bengal Caste

The Garden Transformed: Prince Edward Island, 1945-1980

(V. Smitheram and D. Milne, co-editors)

Community and Agriculture in Two Indian Villages

(G.R. Madan, co-author)

Villages Upward Bound: Community Structure and Technological Development in Selected Indian Villages

(H.F. Kaufman and A. Singh, co-authors)

Eadpur: A West Bengal Village

(S.P. Bose and T.K. Biswas, co-authors)

Contents

Tables

Foreword

As change assails Canada's rural society the perception has spread that urbanization has engulfed our countryside leaving behind a society that is neither rural nor urban. This notion, nourished by lack of information, or an oversupply of misinformation, engenders attitudes of despair among those who care about rural life and indifference among those who could counteract or redirect present trends: journalists, politicians, teachers, clergy, health workers and, most sadly of all, rural people themselves. The publication of a comprehensive text book on Rural Canada is a rare event and, because of the scarcity of such books, a welcome one. Not that scholars and journalists have been silent, indeed there is an extensive literature that goes back to writing in the early nineteenth century and continues up to the present. Those of us who worked to establish the Rural Canada Bibliography have found thousands of titles to list on a new data base. Yet there are few books that give us an overview of the new rural society that is emerging in Canada.

An introduction to Dr. Satadal Dasgupta, the author of this book, is appropriate for those who are unfamiliar with his work. Born in India he attended the University of Calcutta where he earned a Ph.D. in social anthropology in 1964. He then proceeded to the United States as he was awarded a Fulbright post doctoral fellowship at Mississippi State University where he continued to teach as Assistant Professor. He then accepted an appointment to the Department of Sociology and Anthropology at the University of Prince Edward Island in 1969, where he continues to write and teach, now as chair of that department. In the interval he has been active on campus and in the provinces besides producing a scholarly output of books, papers and monographs culminating in this volume which is a revision and updating of a book by the same title issued by the same publishers in 1988.

What to me is an impressive feature of his distinguished career has been the regular contact he has maintained with his motherland, consisting of frequent visits, lectureships and research papers. It seems to me this dual focus on two vastly different cultures has equipped him to develop a holistic view of Rural Canada. In a sense he is an anthropologist who has adopted the rural people of Canada as his tribe. In this case however he has had to depend on a very large number of books and reports to provide the data he needed to reach his conclusions. In fact the sheer volume such references cited is impressive.

Both volumes follow the same outline. In this volume the statistics and bibliography are brought up to date along with reference to studies published since 1988. The revised volume is also enriched with fresh observations and insights. Dasgupta builds his study around the familiar rural/urban continuum a concept some find no longer valid but it serves his purpose to provide a framework of reference for the understanding of the rapid invasion of urban values and global institutions into rural society. He deals on the impact of change on agriculture.

Besides dealing with far reaching changes in agriculture he reviews population figures up to 1991 along a wide range of categories. This is followed by examination of the emerging rural social structure under following topics — community, family, church, school, and local government. He then anticipates the new society that is emerging in his last two chapters but he rightly avoids speculation. The scope of his inquiry is so wide and complex that he has had to depend on the research of others. Since scholarly research and its publication necessarily lags behind the ongoing drama of rapid change which is being reported, it therefore follows that more time is needed for those who must fit the pieces together to give us the big picture of what IS rural Canada.

Looking back to the 1930's when an undergraduate at the University of Toronto I was hungry for a book like this to help me to understand the changes that I had witnessed in my travels to other parts of Canada and had experienced first hand my own rural community. There was still more frustration when as a graduate student at Columbia and Michigan State there were texts like this one prescribed for study about a rural reality in the United States. Of course much of what was described was similar to our own in so many ways and in so many others so misleadingly different.

It has been a long wait for a book like this. This book will help to sharpen an awareness of the extent of change in rural Canada. We rural people, we the Canadian public as a whole need this awareness. The cost of the losses to the land, the cost of the snuffing out of cherished rural values will be widely shared even though the substantial benefits have not. With that awareness will come a greater appreciation of books like this and a demand, one hopes, for many more like it to follow.

R. Alex Sim
Guelph, Ontario

Preface

This volume is a revised and updated version of *Rural Canada: Structure and Change* published in 1988. The goals of this edition have been to update all the statistical data and incorporate the findings and generalizations of the studies on rural Canada published since 1988. While the organization of the book and the chapter titles have remained the same, significant portions of each chapter have been revised and rewritten in light of the new statistical data and substantive materials. As was done in the earlier edition, most of the statistical information used in the present volume have come from the decennial census data. The new statistical data included in the present edition have generally been obtained from the 1991 Canadian census although information from the 1996 census were also used where available and considered appropriate.

Several new volumes on the Canadian rural life were published immediately before and after the publication of the first edition of *Rural Canada* in 1988. Alex Sim's (1988) *Land and Community: Crisis in Canada's Countryside*, for example, was published immediately before the publication of the earlier edition of the present book. Although Sim characterized his book as "a personal statement rather than a sociological treatise" on the crisis in Canadian rural life and community, it was nevertheless a valuable sociological contribution to the understanding of the problems of the contemporary rural Canadian society. An edited volume by Gurcharan Singh Basran and Davis A. Hay (1988), *The Political Economy of Agriculture in Western Canada*, was also published in the same year focussing on agriculture in the western region of the country. Three other edited volumes dealing with the rural society in Canada were published in the 1990s: David Hay and Gurcharan Basran's (1992) *Rural Sociology in Canada*, Ray D. Bollman's (1992) *Rural and Small Town*

Canada; and Richard C. Round's (1997) *Changing Rural Institutions*. It is obviously good news that there is a growing interest in the sociological study of the rural Canadian society as the increasing number of books on the subject appears to attest. However, *Rural Canada* continues to exist so far as the only textbook providing a comprehensive overview of the changing rural life in Canada.

The present edition of the book has a new subtitle, "a rural sociological perspective." The same perspective was used in the earlier edition of the book as well although it was not specified as such. Some rural sociologists in Canada label the rural sociological perspective used in *Rural Canada* as "structural-functional." Indeed, in its evolution as a subdiscipline of sociology in the United States, rural sociology followed this dominant theoretical approach of the time in sociology. The vast majority of rural sociologists continue to follow the structural-functional approach not only in the United States but also elsewhere even though some prefer to follow the emerging "critical perspective" in rural sociology. The structural-functional approach is used in this volume because it is most widely followed in rural sociology both in teaching and research, and the author of the present volume is most conversant and comfortable with it in providing an analysis of the changes in the Canadian rural society. A textbook following the critical perspective in rural sociology is best left to be written by those who consider it to be, for whatever might be the reason, the most "distinctively Canadian" perspective in rural sociology.

Several people provided me with assistance to make the second edition of *Rural Canada* possible. Shawn Carr, Michael Drake and Sherry Doherty, students of my course on Canadian rural society, helped respectively in updating some of the statistical information on rural church and religion, changing agriculture, and rural population. Cynthia McKenna, a former student and a reference librarian at the University of Prince Edward Island was very kind and helpful in finding government reports and documents relevant for my research. Rachel Irving, a student and research assistant, helped in editing and proofreading a number of chapters. My thanks to Judi Wagner, our departmental secretary, who provided prompt and unfailing assistance in typing several versions of the manuscript with great patience and care. Finally, I express my deep gratitude to R. Alex Sim — co-author of the pioneering and now classic study of the Canadian suburbia, *Crestwood Heights* and, more recently, the author of a book on contemporary rural communities, *Land and Community: Crisis in Canadian Countryside* — for honoring me by writing the Foreword for this book.

Satadal Dasgupta
Charlottetown
University of Prince Edward Island

Preface to the First Edition

This book provides a social scientific overview of rural social structure and change in Canada. Although there has been a significant amount of research on Canadian rural life, no attempt has yet been made to put together the findings and generalizations of those studies in a single volume. John MacDougal's (1973) *Rural Life in Canada* was originally published in 1913 and deals with, from a Presbyterian clergyman's point of view, the church and farm problems of that era. Anderson's (1977) *Remembering the Farm* is a nostalgic reconstruction of the traditional rural life in Ontario. The book, *Rural Canada in Transition*, edited by Tremblay and Anderson (1968), includes an excellent collection of papers presented at a conference of the Agricultural Economics Research Council which focus primarily on the agricultural and economic aspects of rural Canada. No book is available which provides a comprehensive overview of the changing rural life in Canada. The present volume is a modest attempt to fill that void.

Since this book presents a synthesis of findings and generalizations of selected studies on rural Canada conducted over the past fifty years, it has drawn its material from a great multiplicity of sources. While these sources have been listed under "references" at the end of this book, I wish to especially acknowledge my indebtedness to the following studies from which I draw extensively: Abramson (1968) on the adjustment of rural migrants to urban centers; Abell (1968) on the mechanization of agriculture; Task force on Canadian Agriculture (1977) on changing agriculture; Gunn (1978) on part-time farming; MacPherson (1982), Cooperative College of Canada (1982), and Hill (1976) on the history, organization and growth of cooperatives; Wood (1975) and Griezic (1975) on the history of farmers' movements; Laskin (1968) on the typology of trade centers; Burnett (1951) on the rural village and small town; Cochrane (1981) on the one-room school; and Rowat (1975), Crawford (1958), and Tindal and Tindal (1979) on the history, typology, and problems of local governments. The statistical information used in this book came mostly from the Canadian census data.

The book grew over several years as I wrote and used several versions of the manuscript to teach a course on Canadian rural society. Although this volume does not have all the features which characterize a textbook, I feel that it can be used as one by those who wish to do so. The book, however, is aimed at a great variety of readers. Care has been taken not to make this book overly technical so that readers with little or no social science background will not find it difficult to read and will be able to use the information presented for their own purposes. Every social scientific concept used in this book has been defined in simple terms without diluting its scientific meaning. The first chapter, for example, attempts to familiarize the reader with the conceptual terms and frame of reference used in this book.

Finally, I would like to thank those who helped me with this effort. Professors Gerald Globetti (University of Alabama), Robert Kaill (Dalhousie University), and Walter Fobes and Terry Pratt (University of Prince Edward Island) read selected chapters and provided useful suggestions and encouragement. Judi MacKinnon, our departmental secretary, typed innumerable versions of the manuscript with great patience and care. Michael McCarthy, Patrick Campbell, and my daughter, Kaberi Dasgupta helped in editing and proofreading. The Senate Committee on Research and Bonnie Suen of the Computer Center at the University of Prince Edward Island respectively provided the funds and expertise in phototypesetting of the manuscript. Lastly, my thanks to my wife, Krishna, whose patience and encouragement allowed the book to become a reality.

CHAPTER ONE

Rural and Urban: Frame of Reference

Like most industrialized countries in the world, Canada today is highly urbanized. The proportion of the rural population in Canada declined consistently during the one hundred years between 1871 and 1971. While in 1871, when the first Canadian census was taken, almost four-fifths of the Canadian population was rural, in 1971 it had declined to less than one-fourth of the total. In 1981, for the first time in Canada's history, the rural segment of the population showed a slightly higher proportional increase than the urban. In 1991, however, the trend reverted back to the higher proportional increase of the urban population compared with the rural.

The picture is essentially the same in all industrialized countries of the world. A little over one-fourth of the population lives in rural areas of the United States. Less than one-third of the population of England and Wales may be considered rural. Only about one-fourth of the economically active male labor force are engaged in farm-related occupations in most of the industrialized and urbanized countries. The massive movement of rural people to cities and the increase of urban populations are evident in the nonindustrialized countries as well. The urban population is growing so rapidly throughout the world that according to one estimate (Weissman, 1967: 44) only about one-third of the world's population was to remain rural by the turn of the twentieth century.

Urbanization, however, has taken clearly different patterns in the industrial and nonindustrial countries of the world (Davis, 1965). In the industrial countries, the process of urbanization started out slowly at the early stage, but once it gained momentum it proceeded at a highly accelerated rate until the urban population exceeded fifty percent of the total population after which it slowed down because there was no longer enough rural population left in those countries to ensure a significant rural to urban migration. The nonindustrial countries, on the other and, are urbanizing at a faster rate today than did the industrial countries during their period of rapid urbanization. But the high rate of urbanization has not changed the proportions of rural and urban populations in the nonindustrial countries to any significant extent. The rate of population growth is so high in these countries that the part of the rural population which migrates to cities is almost entirely replaced by a newly born population within a short period of time. This ensures a continuous stream of rural to urban migration without significantly altering the rural-urban population ratio. Thus, while rural populations in the industrial countries are declining or approaching a stationary level, those of the nonindustrial countries continue to be large even in the face of rapid urban growth.

The world, in other words, will continue to be predominantly rural even though the rural populations of the industrial countries have suffered dramatic depletion. Well over half the population of Asia and Africa is estimated to account for almost three-fourths of the world's population, and to live and work in rural areas at the end of the twentieth century (Jones, 1973: 5). The picture is not going to change significantly, at least for some time to come, notwithstanding Weissman's (1967: 44) prediction to the contrary. This should be the reason enough for a greater depth and rigor in the investigation of rural society in general (Halpern, 1967). A systematic knowledge about rural societies is no less important for the industrial countries. Many urban dwellers of these countries are first-generation urbanites with persistently rural values, beliefs, and outlooks. Indeed, some sociologists (Smith and Zopf, 1970: 496) foresee "a greater ruralization of the city and lesser urbanization of the country" as more and more people move from cities to suburbs and rural-urban fringe areas, and the formation of a "rurbanized" society in the future.

The purpose of this book is to provide a systematic description and analysis of the rural social structure and change in Canada in demographic, economic, and social terms. Such an analysis, however, cannot be made in a conceptual vacuum. The meaning of the factual data can be understood only when they are viewed against a conceptual framework, which is essentially a theoretical "model" based on defined concepts, elements, and their interrelationships. It not only helps to understand the meaning of acquired facts but also to organize them in a systematic manner.

The "rural-urban continuum" is one such model which has been used by both sociologists and anthropologists in analyzing rural and urban societies, even though it has been criticized by some scholars for its various shortcomings. It is basically a comparative model describing the characteristics which typify and distinguish a "rural" culture or society from an "urban." Since the "rural-urban continuum" model is used as a conceptual framework for the present analysis, the following section deals with its origin and form in a historical perspective. The final section of this chapter provides a description of the plan of this book.

DEFINING "RURAL"

The terms "rural" and "urban" are often used as though they are mutually exclusive and easy to distinguish. The use of the two terms as simple dichotomies is seen even in scholarly treatises. But, as we will note in the following, the definitions of "rural" and "urban" are "fraught with pitfalls and uncertainties" (G.J. Lewis, 1979: 21).

Rural and Urban As a Dichotomy

Census reports make a dichotomous distinction between rural and urban. In Canada, for example, communities with a population size of less than 1000 people are classified as rural and those with a population of 1000 or more are designated as urban. Such a dichotomy between rural and urban based on a cut-off point of population size is obviously arbitrary. The arbitrariness of census definitions is further emphasized by the fact that they vary from one country to another. In the United States, for example, the cut-off point is 2500; in Greece, 10,000; and in some Scandinavian countries it is as low as 200. In India, on the other hand, a community will have to have a population of at least 5000 before it will be designated as urban.

The general population size of a country and its cultural norms appear to determine the census definition it adopts (Jones, 1973: 6-7). Countries with small and dispersed populations consider small nucleated settlements as urban. Where rural populations are large and agriculturists live in nucleated settlements, even large villages are designated as rural unless they satisfy other criteria of urban such as population density and occupation. The wide variation in census definitions of rural and urban obviously makes cross-societal comparisons extremely difficult.

There is also a long tradition both in sociology and anthropology of describing rural and urban as dichotomous "ideal types" stressing the dissimilarities and discontinuities between them. These ideal type constructs are not generalizations based on research but are conceptual models describing essential qualities of contrasting social systems in exaggerated terms without specifying the details which might be observed in reality. They are useful in classifying real societies as rural and urban in order of resemblances to them.

Some of the well-known ideal type constructs are Tonnies's (1887) *Gemeinschaft* and *Gesellschaft*, Durkheim's (1893) *Mechanical* and *Organic Solidarity*, Cooley's (1909) *Primary* and *Secondary Groups*, and Becker's (1950) *Sacred* and *Secular Societies*. Each of these authors attempted to articulate the fundamental differences between pre-urban and urban social systems as contrasting dichotomous types. Although these ideal types vary in emphases on specific dimensions of the contrasting social systems, they share a basic underlying similarity in that they all deal with the same generic aspect of the two opposing social types. Each emphasizes somewhat different aspects of what are fundamentally the same types of groups.

Gemeinschaft and Gesellschaft

The distinction between *Gemeinschaft* and *Gesellschaft* is based essentially on the relationships growing out of fundamental differences in the normative expectations of the two contrasting types of social organizations. Tonnies maintained that social relationships rest on either of two opposite principles: natural will and rational will. In *Gemeinschaft* societies, where the natural will predominates, social relationships are based on understanding, unity, and sentiment. Obligations between persons are diffuse; means and ends are not clearly distinguished; and social control is informal and is guided by deeply internalized attitudes of love and respect. By *Gemeinschaft* Tonnies meant a "social order bounded together by a unity of wills" in which "family and social institutions naturally created cooperation" (Liebersohn, 1988: 7).

Rational will takes precedence in *Gesellschaft* societies where social relationships are based on rationality and deliberate calculation of appropriateness of means to ends. Rationality, formality, and specificity form the fundamental principles of social interaction. Social control in these societies is formal and directed by institutionalized authority. Tonnies believed that the rural village and town correspond most clearly to the *Gemeinschaft* relationships. The city and metropolis, on the other end, exemplify the *Gesellschaft* relationships. Tonnies thus conceptualized two contrasting social systems based on two opposing ways of life. "The city, by virtue of the primacy given to commerce encourages *gesellschaftlich* relations; the country, the village and the town by virtue of the primacy given to family and history, give rise to *gemeinschaftlich* relations" (Bonner, 1997: 21).

Mechanical and Organic Solidarity

Durkheim's conceptualization of pre-urban and urban societal types is similar to that of Tonnies in that he also contrasted them in terms of their relational characteristics by designating the former as "mechanically solidary societies" and the latter as "organically solidary societies." Durkheim saw the mechanically solidary societies as "segmental" entities built of repetitions of family-like groups alike in beliefs and conduct, and integrated together by a moral and mental homogeneity. A totality of beliefs and sentiments which Durkheim called the "conscience collective" is shared by all members. Offence against the collective conscience in these societies is considered a "moral" offence punishable by "repressive" law for the protection of common interest (McKinney and Loomis, 1963: 13).

The organically solidary societies are "organized" entities built of differential and functionally specialized parts whose mutual interdependence holds these societies together into functioning wholes. The increase in division and specialization of labor in these societies gives rise to individualism and heterogeneity in beliefs, conducts, morality, and sentiment. The moral and structural diversification of the society weakens the collective conscience to the point that it ceases to be a basis for societal solidarity. Crime is no longer an offence against moral sentiments rather against personal "rights," and is dealt with "restitutive" law to preserve cooperative relations among persons with different interests (McKinney and Loomis, 1963: 13).

Primary and Secondary Groups

Cooley, a social psychologist, was more interested in the essential characteristics of small groups and their importance in social relationships rather than in the dichotomous character of societal types. He perceived a crucial difference between what he called "primary" and "secondary" groups in their characteristic social relationships. Cooley's conception of the "primary group" is, however, very similar to Tonnies's notion of the *Gemeinschaft*. The primary groups are small in size with intimate, face-to-face, and continuous cooperation among members lasting over a long period of time. Relationships within the primary group are personal, inclusive, and spontaneous; members interact in an enduring relationship as total persons to attain commonly shared collective goals; and fuse their individualities in a strong "we" feeling. The integration in a primary group is achieved through a reliance on informal social control. They are "primary" because they are "the primary phase of society," and are most likely to occur within a family, peer group, neighborhood or small community.

Although Cooley himself never used or explicitly discussed the notion of the "secondary group," he conceptualized it as the polar opposite of the primary group by implication. The secondary group then, by implication, is large in size with formal interaction among its members, goal-specific, and of relatively short duration. Because of the plurality and disparity of ends of its members, the secondary group tends to encourage competition rather than cooperation. Its relationships are impersonal, and perceived by its members as means to attain ends rather than being ends in themselves. The integration of a secondary group is based not on a strong "we" feeling but on formal structural principles. The secondary groups predominate in a large town, city, or metropolis. An urban community, primarily because of its large population size and high density, is characterized by formal, impersonal, goal-specific, and exclusive relationships among its residents.

Sacred and Secular Societies

The fundamental distinction between Becker's "sacred" and "secular" societal types is that in the former tradition and ritual play a large part in an individual's life while in the latter the influence of tradition and ritual is minimal and rationality and science are pervasively dominant. The geographic, social, and mental isolation of the people in a sacred society gives rise to the "fixation of habit and neophobia, relations of avoidance, and traditional in-group — out-group attitudes" (McKinley and Loomis, 1963: 16). The sacred society is also characterized by a minimal division of labor, strong kinship and family ties, an impermeable value system, and sacred sanctions for all forms of activity including economic.

The secular society, on the other hand, is accessible in geographic, social, and mental terms which makes habit fixation difficult. Change and progress are idealized, and innovation is frequent in these societies. The distinctive structural features of the secular society are an increased division of labor, weak kinship organization, permeable value system, dominance of rational behavior, and formal legal sanctions against social deviance.

Common Features of the Societal Typologies

Although none of the above authors used the terms "rural" and "urban" in labelling their societal types, they all attempted, at least by implication, to articulate the foundations of pre-urban and urban social orders in contrasting terms. Their models specify a dichotomy of society stressing the differences and discontinuities between the rural and the urban. They, however, emphasized somewhat different aspects of fundamentally the same types of groups. A society which is *Gemeinschaft* to Tonnies is viewed by Durkheim to be integrated on a principle of "mechanical solidarity" in which, according to Cooley, "primary groups" dominate and the society as a whole, from Becker's viewpoint, is oriented in traditional and "sacred" terms. In contrast, the *Gesellschaft* society is based on "organic solidarity" with a predominance of "secondary group" relations and an overall "secular" orientation.

The differences these authors emphasized in contrasting the rural and urban societal types can be summarized in terms of three dimensions (Ryan, 1969: 310): homogeneity-heterogeneity; primariness-secondariness; and sacredness-secularity. Both Tonnies and Durkheim emphasized the homogeneity-heterogeneity dimension or the degree of social differentiation. Cooley focused on the differences in the primariness-secondariness aspects in group relations. The emphasis on the sacredness-secularity in group relations was obviously central to Becker's typology. In sum, Tonnies, Durkheim, Cooley, Becker and others were describing different aspects of a single construct typology consistent with "pre-urban" (or "rural") communities as contrasted with "urban" communities.

Another common feature of this typological tradition is that the foundations of contrasting social orders are described in dichotomous terms emphasizing dissimilarities and discontinuities between two polar types. Each conceptual type is expected to serve as a theoretical yardstick to classify real communities or societies as "rural" or "urban." But empirical entities show a wide range of variation in their degree of conformity to one conceptual type or other. No two communities show as sharp a discontinuity as a "rural" and an "urban" community as implied by these typologies. Thus, some sociologists (Duncan and Reiss, Jr., 1956: chap. 2) argued that such a dichotomous conceptualization of the rural and urban is fallacious, and hardly exists empirically as such. They are, however, useful if considered as representing the ideal polar types of the "rural-urban continuum." Even though no community or society would exactly replicate either of the polar types, they could be placed somewhere along the continuum according to the degree to which they correspond with one polar type or the other.

Folk-Urban Continuum

Robert Redfield's (1947) "folk-urban continuum" sought to achieve a greater sophistication than the earlier societal typologies by utilizing traits and variables that were of a general and more abstract nature, and emphasizing the transformation which occurs from one pole to another. In other words, this typology is also a theory of social change which can be used to identify the nature and the direction of the social processes involved. Redfield initially described only the folk end of his schema at length by linking together a set of ideal attributes, and implied that the urban end was simply the polar antithesis of the folk.

Redfield's folk society is a small, isolated, nonliterate and homogeneous grouping of people with a strong sense of solidarity. The technology of the folk society is simple with no basis of division of labor except sex which makes the group economically independent of other groups. The ways of living is conventionalized into a coherent system: a culture. Behavior among members is traditional, spontaneous, uncritical, and personal, and there is no legislation of habit or motivation for experimentation or reflection. Kinship is central to all experience with the family being the unit of action. The sacred prevails over the secular, and all activities, including economic, are considered ends in themselves.

According to Redfield (1953), the "moral order" as opposed to the "technical order" is predominant in the folk society. By the moral order he meant those aspects of life and culture which are sacred, normatively impermeable, and true and right by the fact of their being. The technical order, which dominates the urban society, includes those aspects of culture which are instrumental in valuation, and those ways and things which are valued not for the traditional rightness but for efficacy in accomplishing some ends.

Redfield's folk society is similar to Tonnies's *Gemeinschaft* or Durkheim's Mechanically Solidary society. His distinction between "moral" and "technical" orders is not significantly different from Tonnies's "natural" and "rational" wills or Becker's "sacred" and "secular." Indeed, Redfield (1943: 68-71) himself indicated his indebtedness to many of the earlier societal typologies, especially those of Tonnies and Durkheim, and that his folk society is a restatement of several of these earlier conceptualizations based on his research of primitive societies. Redfield's unique contribution lies in the fact that he utilized the typological conceptualizations in organizing and interpreting his empirical materials, and empirically identified communities which represented points along his folk-urban continuum (Redfield, 1941).

Urbanism As a Way of Life

Redfield did not describe the urban end of his continuum in any great detail except by implying that it was the polar antithesis of the folk society. It is in Wirth's (1938) classic article, "Urbanism as a Way of Life," that the urban end has been dealt with great elaboration as a conceptual type fundamentally different from a folk society.

Louis Wirth suggested that the social characteristics of the urban life derive from its large size and high density of population, and from its social heterogeneity. Secondary, rather than primary, group relations predominate in urban society which results in anonymity, and in impersonal, transitory, superficial, and utilitarian social contacts. Because of high division of labor and a greater interdependence of specialists, a ramified and sharpened, although more flexible and mobile, system of social stratification based on income and social status comes into existence. Symbols and stereotypes become important because of lack of personal acquaintance, and people become involuntarily segregated by race, language, income, and class groups. Kinship bonds become weakened and the social significance of the family declines contributing to the loss of traditional bases of solidarity. Territorially based social units, such as neighborhoods, become replaced by varieties of interest groups.

Wirth (1938: 11) emphasized that "the bonds of kinship, of neighborliness, and the sentiments arising out of living together for generations under a common folk tradition are likely to be absent, or at best, relatively weak in an aggregate, the members of which have such diverse origins and backgrounds. Under such circumstances competition and formal control mechanisms furnish the substitutes for the bonds of solidarity that are relied upon to hold a folk society together."

Orthogenetic and Heterogenetic Cities

In his later writings, Redfield (Redfield and Singer, 1971) differentiated between two types of cities: orthogenetic or preindustrial, and heterogenetic or industrial, products of primary and secondary urbanizations respectively. Revising his original thesis of two stages of societal development, folk and urban, Redfield later identified three evolutionary stages: folk, peasant, and urban. Although he originally designated both primitive and peasant societies as "folk societies," he now differentiated between the two.

The "folk society," according to Redfield, came to be exemplified by the "precivilized" and preliterate bands of nomadic hunters and gatherers of the preneolithic times the remnants of whom continue to exist in different parts of the world even at the present time. The "peasant society" evolved after the agricultural revolution of the neolithic times as settled communities of agriculturists came into existence along with the preindustrial "orthogenetic" cities which served as the religious and political centers for the peasant countrymen around them. Redfield called this phase of urbanization "primary" because the peasant society and the correlated urban center together constituted a culturally continuous entity based on the "moral order" which characterized the folk society as well.

Technical development and increasing contacts with other cultures led to the secondary urbanization and development of the secular, "heterogenetic" city based on the "technical order." The "urban way of life" described in the characterizations by Wirth (1938) and others is principally a consequence of secondary urbanization. While the city and country are more closely integrated in the primary phase of urbanization because they rest on a common culture, they cannot be integrated on a common cultural consciousness in the secondary phase of urbanization, for, according to Redfield, "there is none." Mutuality of interests and "symbiotic" relations constitute the only bases of rural-urban integration in this phase of urbanization along with the development of a consciousness of cultural differences and mutual interdependence: the city being a "service station" for the country, and the country a "food basket" for the city.

Redfield thus expanded his "folk-urban continuum" into "folk-peasant-secondary urban continuum" in his later writings. While the "folk" refers to preliterate, primitive communities, the referent of the "peasant" is the agricultural countryman integrated socially, economically, and politically, and sharing a common culture with an orthogenetic urban center. It is the heterogenetic city, rather than the orthogenetic, a product of the secondary urbanization, which represents the "urban" end of the continuum.

Referent of "Rural"

The referent of the term "rural," as opposed to the "urban," thus varies according to the stage and level of urbanization of a society. The "rural" refers to "peasants" in societies where urbanization prevails at a primary level because of its relative isolation and a low level of technical development. Most societies in Asia, Africa, and South and Central America, and a few even in Europe, are still at a primary level of urbanization and only recently have started to industrialize. The rural residents of these societies account for the overwhelming majority of their total populations, and are primarily peasants.

Although both rural and agricultural, peasants constitute a distinct social type different from both the "primitive folk" and the rural farmers of industrial societies. They are different from the primitive folk in that they do not exist as isolated autonomous units but always form a segment of a larger entity of which the orthogenetic urban center is a part. They are "definitely rural — yet live in relation to market towns; they form a class segment of a larger population which usually contains also urban centers ... They lack the isolation, the political autonomy, and the self-sufficiency of tribal populations; but their local units retain much of their old identity, integration, and attachment to soil and cults" (Kroeber, 1948: 284). On the other hand, peasants are different from rural farmers of industrial societies in that as agriculturists they aim at subsistence rather than reinvestment. Unlike commercial farmers in Canada or the United States, peasants use their resources to grow crops and raise livestocks primarily to satisfy the needs of their households instead of reinvesting the products of their labor to obtain a profit. Although they exchange some of their surplus with others for goods, services, or resources, peasants are rarely oriented toward a market economy, fluctuating prices, and supply and demand. In sum, peasants as a social type are part of a larger social structure of which a market town or a pre-industrial city is also a part. They are subsistence agriculturists producing primarily for their own consumption and also to support the urban segment of their total social structure. The urban dwellers in a peasant society "depend upon peasants both for their food and for any income that may accrue to them" (Wolf, 1966: 12).

In an industrial society, however, the "rural" does not refer to "peasants." A rural farmer in such a society is distinct from a peasant in economic, technological, and social terms. The farmers in an industrial society are commercially oriented rural villagers who produce food and other articles not for consumption by their families but to make a profit. Their productive activities respond to supply and demand, and fluctuating prices in the market. The farms they own are large, heavily capitalized, and scientifically operated. “If there are any peasants left in such societies, they occupy a secondary position in the creation of wealth” (Wolf, 1966: 12). The urban dwellers in an industrial society do not depend for food on peasants who work small units of land with traditional techniques but on large commercial farmers who operate “factories in the field” applying the new industrial technology. The integration of the rural segment of an industrial society with the heterogenetic city is not based on sharing of a common culture but on a consciousness of cultural differences and symbiotic interdependence. But the secondary urbanization in an industrial society “produces not only a new form of urban life in some part in conflict with local folk cultures but also new social types in both city and country. In the city appear ‘marginal’ and ‘cosmopolitan’ men and an ‘intelligentsia’; in the country various types of marginal folk: enslaved-, minority-, imperialized-, transplanted-, remade-, quasi-folk, etc., depending on the kind of relation to the urban center” (Redfield and Singer, 1971: 345).

Thus, in a traditional society which is predominantly at a primary phase of urbanization, and where the secondary urbanization is only at a preliminary stage, the transition from “rural” to “urban” involves a transformation from a peasant society to a society with secondary urbanization. In an industrial society such as Canada, however, the rural society has moved far from the peasant stage, which probably existed prior to the introduction of elaborate machinery, toward the secondary-urban end of the continuum. The postpeasant rural communities of a modern society are far more commercial than their peasant counterparts of a traditional society, and are oriented toward a market economy.

Redfield’s folk-peasant-secondary urban continuum is a typology as well as a process describing evolutionary changes in human societies. Redfield not only delineated polar types and described their outer limits, especially those of the folk (1947) and of the peasant (1956), but also the process as a society goes through the transition from folk to urban (Redfield, 1941; Redfield and Singer, 1971). Although, as will be shown in the following section, his theory has been severely criticized by several sociologists and anthropologists, Redfield's folk-peasant-secondary urban continuum continues to be widely used to explain rural social change.

Rural-Urban Continuum

While Redfield's folk-peasant-secondary urban continuum dealt with historic change of communities in the evolutionary context, Sorokin and Zimmerman's (1929) rural-urban continuum dealt with historic change not so much in the context of evolution as in the analytical contrast between rural and urban life. They maintained that the rural and urban societies are individually characterized by a distinctive set of causally connected traits, and that the agricultural occupation, rather than the population size and density, primarily differentiates the rural from the urban. "As soon as one takes the agricultural occupation and the people engaged in it, he finds other differences enumerated." The "rural world" is thus composed of a totality of cultivators and their families which in turn induces a predominance of nature over the anthropo-social environment, the smallness of community size, a homogeneity and a low density of population, low social differentiation and stratification, less intensive social mobility, and a predominance of personal, primary, and durable social relations. The nonagricultural occupations of the residents of the "urban world" isolate them from nature in a man-made environment where the size of the community is large, population density and heterogeneity is high, social differentiation and stratification, and mobility are intense, and social relations are impersonal, formal, and short-lived. The transition from a "purely rural community to an urban one" is gradual with each step being associated with a proportionately decreasing agricultural population. Sorokin and Zimmerman maintained that "many differential characteristics of the rural and urban community would consist not so much in the presence of certain traits in rural, and their absence in urban communities, as much as in a quantitative increase or decrease of these characteristics."

Real but Relatively Unimportant?

Several writers criticized the rural-urban continuum for its shortcomings as a model to study social change in rural and urban societies. Hauser (1967: 514), for example, dismisses the various dichotomies and continuua, including the rural-urban, as "catchy neologisms which get confused with knowledge"; Oscar Lewis (1967: 491) thinks that "their heuristic value as research tools has never been proven"; and Hofstee (1960: 105) brands them as "vulgar Tonniesim" uncritically glorifying the "old-fashioned rural life."

Perhaps the most serious criticism against the rural-urban continuum relates to its alleged demographic and ecological determinism. Oscar Lewis, one of its earliest critics, observed (1967: 497) that "the variables of number, density, and heterogeneity as used by Wirth are not crucial determinants of social life or personality," and that "there are many intervening variables" which affect social life both in the country and the city. "Primary groups are just as important psychologically for city people as they are for country people"; although the secondary relations are much greater in the city than in the country, "their psychological consequences are minor." Warm and intimate relations are not "necessarily or exclusively a function of frequency of contact and fewness of numbers" but "are influenced by cultural traditions." There are perhaps "deeper, more mature human relationships among sympathetic, highly educated, cosmopolitan individuals who have chosen each other in friendship, than are possible among sorcery-ridden, superstitious, ignorant peasants, who are daily thrown together because of kinship or residential proximity."

In his study of a Guatemalan folk society, Tax (as quoted in Redfield, 1947: 308) observed "relationships impersonal, with formal institutions dictating the acts of individuals, and the family organizations weak, with life secularized, and with individuals acting more from economic and other personal advantage than from any deep conviction or thought of the social good." On the other hand, Gans (1962) discovered the "urban villagers" in a working class neighborhood in the center of Boston with a high level of social cohesion based on interwoven kinship networks and primary contact with familiar faces. These ethnographic evidences have prompted some writers to question the validity of the use of population and spatial criteria to distinguish the "rural" from the "urban." Pahl (1968: 293), for example, observes that any "attempt to tie particular patterns of social relationships to specific geographical milieux is a singularly fruitless exercise" and Constandse (1968: 301) agrees with Pahl noting that "a typology of settlements, from small to the large, can never form the basis for a continuum which classifies social structures." Reiss (1955: 43) pursued the same argument suggesting that "empirically, at least, 'urban' can be independent of size and density... large size and high density of settlement are not always conditions for an urban way of life in any given community."

In an excellent article titled, "The Rural-Urban Continuum: Real but Relatively Unimportant," Dewey (1960: 60-6) maintains that "it is not logical to hold that the rural-urban continuum has no universal or general referents, merely because of the mistaken assignment to urbanism of a welter of cultural items which can be, and are, independent of city environments." He further suggests that there is ample evidence to support Wirth's observations that variations in size and density of population induce concomitant variations in five qualities: anonymity, division of labor, heterogeneity, impersonal and formally prescribed relations, and symbols of status which are independent of personal acquaintance. "The logic of custom and of the English language supports the retention of 'rural' and 'urban' to designate the extremes of the continuum defined by the five items."

Dewey argues that anonymity is inevitable for the vast majority of people in an urban center and almost impossible in a rural community. The urban dwellers interact with a large number of people in their daily activities a great majority of which is secondary in nature. But the bulk of the rural dwellers' contact are with members of their families, neighborhoods, and villages. Not only their social contacts are few in number but also mostly primary in nature. A city also maintains a multiplicity of occupational specialties as the degree of its division of labor increases with the increase in population size and density. "Whereas a small rural community can present an undifferentiated occupational pattern, a city of a million most certainly cannot." A minimal degree of heterogeneity is inescapable in large cities, and it increases as the city increases in size. A rural community, on the other hand, is relatively more homogeneous. Life in large urban communities without some formal and explicit rules is unthinkable and certainly does not exist. Dewey observes that good intentions and informality will not suffice to "induce even the minimal order requisite for life in a metropolis, regardless of time and place." Relations in a city are bound to be guided by impersonal and formally prescribed rules while in rural communities absence of anonymity and predominance of face-to-face interaction enables relations to be personal and informal. Finally, in an urban milieu people are identified and ranked by means of overt symbols of status for "the pervasive anonymity of the large city demands some means of identifying the functionaries essential to daily living." In the small, personal, and intimate social world of rural communities, however, overt symbols of status are not essential for daily living. Dewey emphasizes that the importance of the rural-urban continuum rests "in the recognition that the five elements are inevitable accompaniments of urbanization and must be taken into consideration in understanding it," and dismisses most other attributes of rural and urban as "mere cultural correlates of urbanization at a particular time and place." Dewey argues, however, that the rural-urban continuum is of "minor importance for sociology," even though the "influences of rural and urban environments upon social organization and individual behavior will remain important facts."

A Conceptual Framework

The argument that the rural-urban continuum should not be accepted as a generalization based on research, and that its utility can be seen only as an ideal-type construct (Hauser, 1967: 509) is perhaps a valid one, but not the claim that it has outlived its usefulness even in the latter role. Despite the fact that rural-urban continuum has come under severe criticism by both sociologists and anthropologists, it continues to retain its "value as a framework within which to explore and interpret the nature and variations of a defined social system" (Jones, 1973: 12). While a degree of bias in favor of "old-fashioned rural life" was manifest in earlier dichotomous typologies, the latter conceptualizations, especially those of Redfield's, were far more objective and largely based on empirical observations. As long as it is accepted that ideal-type constructs are not mirror images of empirical entities, deviations from them, as observed by Tax, Lewis, Gans and others, are to be expected and do not necessarily negate their heuristic values. Indeed, by definition ideal-type constructs are hypothetical, simplified, and somewhat exaggerated conceptualizations of social systems describing their essential qualities without specifying the details and deviations which might be observed in real situations. It is against these ideal types that "real social situations could be assessed according to the extent or relative degree to which they resemble either pole" (Jones, 1973: 12).

The rural-urban continuum can indeed be used "as a basis for classifying, or as a framework for discussing, recent trends" and as an "indicator for future social change (the continuum representing a process), or for both purposes" (Jones, 1973: 12). Lupri (1968: 300) also makes the same argument in stating that the rural-urban continuum is "a conceptual framework within which certain rural-urban differences may be meaningfully interpreted." It should be emphasized, however, that the polar extremes of the continuum are not constant, and the relevance and relative dominance of the characteristics included may vary in different cultural situations. The transformation occurring along the continuum is also not necessarily continuous or lineal. Above all, as Jones (1973: 12) points out, "the concept of the continuum allows one to embrace the idea that any social system can be classified in relation to others, and that its changing nature, as a result of alterations in its characteristics, is not only complex but can also be comprehended in relation to changes in other systems and the relative changes occurring in its own various characteristics."

THE TRANSFORMATION OF RURAL CANADA

The social change that rural Canada has undergone over the last one hundred years or so can be analyzed and meaningfully interpreted within the conceptual framework of the rural-urban continuum. Although some writers claim that North American farmers were not peasants "even prior to the introduction of elaborate machinery" (Foster, 1967: 7), the sociological descriptions of the early rural communities of Canada show that they closely resembled peasant communities as small, relatively isolated, and economically self-sufficient units with kinship and religion providing the fundamental bases of their social organization (Miner, 1939; Fortin, 1964). The technological development, especially in agriculture, as well as increased contact with industrial cities, movement of people from rural communities to urban centers, and spread of urban beliefs and values due to the improved means of transportation and communication have radically transformed rural Canada in recent decades. The transformation has been so dramatic, and urban influence so pervasive, not only in Canada but also in most industrial societies in the Western world, that some scholars have proclaimed the extinction of the rural society or at least of any meaningful distinction between urban and rural society (Bender, 1975: 185).

One significant element of the rural social change in Canada is the migration of rural people to urban centers in large numbers. The technological transformation and the substitution of capital for labor in agriculture contributed to the decline in subsistence farming and increase in rural outmigration. The rural outmigration, in turn, has profoundly affected the rural social structure and institutions. The nature of this demographic transition, its sources of origin, and implications are discussed in Chapter 2.

Chapter 3 deals with the factors which contributed to the technological transformation of agriculture; and its economic and organizational consequences are analyzed in Chapter 4. Indeed, most of the changes in rural population characteristics are related to the changing agricultural patterns. The commercialization and mechanization of agriculture are primarily responsible for the drastic occupational shift in rural areas. Not only has it thrown most of the subsistence farmers out of agriculture but also forced many of those who remain in farming to join cooperatives and farmers' organizations to protect their interest against the growing agrifood industries and corporations.

The changed demographic, economic, and technological circumstances of rural Canada have radically altered the rural social structure and institutions. The traditional rural community has lost its relative economic self-sufficiency and social autonomy, and has been reduced to an "affective" entity satisfying only the social and emotional needs of its residents. The rural family is increasingly resembling the urban in size, composition, internal relationships, and functions. These changes in the rural community and family are discussed in Chapter 5.

Chapter 6 deals with the change and continuity in the rural church, school, and local government. The influence of religion in the rural life has declined significantly with the increasingly secular orientation of a technologically developed rural society. The declining rate of church attendance and the rising cost of maintenance have forced many churches to close their doors. The one-room schools which once dotted the countryside have all but disappeared with school consolidation and a changed attitude of rural people toward formal education. The rural local government is losing its autonomy in decision-making as it increasingly depends on provincial and federal governments for funds to provide modern public facilities and services to its people. The shrinking tax base and increasing demands for more and better services have made the rural local government to act more and more like an appendage of its provincial and federal counterparts.

The final chapter of the book provides a summary of the findings and attempts to relate them to the conceptual framework described in the present chapter.

SUMMARY

Since the rural-urban continuum model is used in this book as a conceptual framework to analyze the rural social change in Canada, its form and development has been presented in this chapter in a historical perspective. Although conceptualized in dichotomous terms in earlier writings, such as those of Tonnies, Durkheim, Cooley, and Becker, the "rural" and "urban" have been viewed as a continuum by Redfield, and Sorokin and Zimmerman. While Redfield's folk-peasant-secondary urban continuum deals with the historic change of communities in an evolutionary context, Sorokin and Zimmerman's rural-urban continuum deals with it in an analytical contrast between the rural and urban. Inspite of the fact that the rural-urban continuum model has been severely criticized by both sociologists and anthropologists for some of its shortcomings, it continues to be widely used as a model to analyze social change. As Bell and Newby (1972: 42) observe, "though put to death by several writers on the subject, it still refuses to lie down"; it is the heuristic appeal of the perspective "of being a relatively simple way to conceptualizing social change and of classifying communities" has enabled it to continue to exert its influence.

CHAPTER TWO

Rural Population: Distribution and Composition

The distinctive patterns of rural and urban life are often explained in terms of their differences in population and occupational characteristics. The predominance of personal and informal relations, primary groups, and kinship in rural areas are a function of the smallness of their community size, a low density of population, and agriculture as an occupational activity. The large size and high density of population, and nonagricultural occupations contribute to impersonal and formal relations, predominance of secondary groups, and weakening of kinship ties in urban communities. Although there is a considerable disagreement among social scientists about the relative significance of population and occupational characteristics as determining factors, there can be no doubt about their importance in influencing the different aspects of rural and urban life. Indeed, many of the social and economic changes in rural Canada can be explained in terms of the recent transformation of agriculture as an occupation and consequent changes in population distribution and composition. The rural to urban movement of people and the changing composition of rural populations have had far reaching consequences for rural life in Canada.

The rural population characteristics are examined in this chapter in terms of distributional and compositional trends. The discussion of distributional trends involves an examination of the patterns of rural-urban migration and their effects on the distribution of people in rural Canada. The analysis of compositional trends deals with the changes in age and sex compositions, marital, occupational, and educational statuses, and ethnic structure.

DISTRIBUTIONAL TRENDS

A major problem of analyzing the distributional trends of rural and urban populations on the basis of the data provided by the decennial censuses is that the census definitions of "rural" and "urban" populations changed several times since 1871 (Kralt, 1976: 3-7). Prior to 1951, the census definitions of "urban" populations included those living in incorporated communities regardless of size while those residing outside of incorporated communities were classified as "rural." The "urban population" was redefined in 1951 as including those people who live in communities of 1000 or more, whether incorporated or not, plus those living in the fringe of a Census Metropolitan Area (i.e., a metropolitan city with a population of 100,000 or more). The same definition was used for the 1956 census as well excepting for the fact that people living in the fringe areas of large cities with a population of 25,000 or more were also included in the "urban population."

The census definition of "urban" was revised again in 1971 which included residents of (1) incorporated communities with a population size of 1000 or more; (2) unincorporated places with a population of 1000 or more and a density of at least 1000 people per square mile; and (3) built up fringes of (1) and (2) having a minimum population of 1000 and a density of 1000 people per square mile. The last definitional change which was made in 1976, and used for the subsequent censuses in 1981, 1986, 1991, and 1996 as well, classified people living in communities, incorporated or not, with a minimum population size of 1000 and a minimum population density of 1000 per square mile (or 400 per square kilometer) as "urban," and people living outside of these communities as "rural."

The changes in rural-urban proportions of populations in different decennial censuses (Table 2.1) are thus partly affected by the definitional changes. The almost seven percent decline of the rural population in 1951 (38.4 percent), for example, from that of 1941 (45.7 percent) was due partly to the change in the definitions of "rural" and "urban" in 1951. If the 1941 definitions were also used for 1951, the proportion of the rural population would have been 43.3 percent rather than 38.4 percent. On the other hand, the definitional change of 1971 had very little effect on the change in rural-urban populations. The proportional trends, however, are not affected as significantly as the actual growth rates of rural-urban populations by definitional changes.

TABLE 2.1

RURAL AND URBAN POPULATIONS OF CANADA BY NUMBER (IN THOUSANDS) AND PERCENT, 1871-1991

Census Year	Canada		Rural		Urban	
	Number	Percent	Number	Percent	Number	Percent
1871	3,689.9	100	2,966.9	80.4	722.3	19.6
1881	4,324.8	100	3,215.3	74.3	1,109.5	25.7
1891	4,833.2	100	3,296.1	68.2	1,537.1	31.8
1901	5,371.3	100	3,357.1	62.5	2,014.2	37.5
1911	7,106.6	100	3,933.7	54.6	3,272.9	45.4
1921	8.787.9	100	4,435.8	50.5	4,352.1	49.5
1931	10,376.8	100	4,804.7	46.3	5,572.1	53.7
1941	11,506.7	100	5,254.2	45.7	6,252.4	54.3
1951	14,009.4	100	5,381.2	38.4	8,628.3	61.6
1961	18,238.2	100	5,537.9	30.4	12,700.4	69.6
1971	21,568.3	100	5,157.5	23.9	16,410.8	76.1
1981	24,343.2	100	5,907.3	24.3	18,435.9	75.7
1991	27,296.6	100	6,390.0	23.4	20,906.9	76.6

Source: Census of Canada.

Rural Decline

During the one hundred year period between 1871 and 1971, the population of Canada changed from four-fifths rural to four-fifths urban (Table 2.1). The proportional decline of the rural population continued consistently in every decade from 1881 to 1971. The rural-urban proportions of the Canadian population became almost identical in 1921. It was in 1931 when the proportion of urban population exceeded that of rural for the first time. The most rapid percentage increase of the urban population took place in the decades between 1941 and 1971.

According to Porter (1961: 137-40), the rural decline in Canada started during the years of "great depression" between 1871 and 1901, when "the high hopes of Confederation and railroad building were not fulfilled," and "more people left the country than entered it." Cities and towns absorbed three-fourths of the total increase in the Canadian population during this period. By 1901, almost one-fourth of the total population of Canada lived in towns of 5,000 or more people.

Porter terms the decade between 1901 and 1911 as the period of "wheat boom." With the discovery of the Canadian prairies and methods to farm the prairie land, and the expansion of the market for the industries of central Canada to the prairies "made all Canada prosperous by its wheat exports." Even during the prosperous period of prairie settlement, the trend toward urbanization continued. The population of Canada increased by two million people in the decade between 1901 and 1911 due to increased immigration from Europe, but only a little over half a million of these people settled in rural areas.

The rural decline continued from 1921 to 1931 but the outmigration from rural areas to towns and cities slowed down during the 1930s due to the disastrous economic depression which was felt equally in urban and rural areas. Farmers had little incentive to join the growing number of unemployeds in the towns and cities.

During the war years in the 1940s, the Canadian economy started to grow again as a response to the war needs and the expanding world market for agricultural products. Although agricultural productivity continued to increase due to mechanization of farms and application of modern and scientific knowledge into farming, the rural people were lured from agricultural occupations by well paying jobs in the cities.

Farm income started to shrink again as the export of agricultural products declined in the 1950s and 1960s with the establishment of the European Common Market and increasing agricultural self-sufficiency of European and other countries.

"Once more the farmers realized that they had lower incomes, worked longer hours, and fewer conveniences than the rest of society. The move from farms accelerated" (Porter, 1961: 140).

Every region of Canada suffered a consistent proportional decline of its rural population in every decade until 1971 excepting British Columbia which showed a slight proportional increase of the rural population in 1921 and 1941 (Table 2.2). However, British Columbia, which had the second highest percentage of rural population among the regions in 1871, suffered the second highest rate of proportional decline of its population by 1991.

TABLE 2.2

PROPORTIONS OF RURAL AND URBAN POPULATIONS OF CANADA BY REGIONS, 1871-1991

Census Year	Atlantic Provinces*		Quebec		Ontario		Prairie Provinces**		British Columbia	
	Rural	Urban	Rural	Urban	Rural	Urban	Rural	Urban	Rural	Urban
1871	88.2	11.8	77.2	22.8	78.0	22.0	96.0	4.0	91.2	8.8
1881	84.4	15.6	72.1	27.9	70.1	29.9	83.5	16.5	81.6	18.4
1891	84.1	15.9	66.4	33.6	61.3	38.7	73.1	26.9	62.0	38.0
1901	75.2	24.8	60.3	39.7	57.1	42.9	75.4	24.6	49.5	50.5
1911	67.9	32.1	59.8	48.2	47.4	52.6	64.7	35.3	48.1	51.9
1921	63.0	37.0	44.0	56.0	41.8	58.2	64.0	36.0	52.8	47.2
1931	62.2	37.8	36.9	63.1	38.9	61.1	62.4	37.6	43.1	56.9
1941	61.5	38.5	36.7	63.3	38.2	61.8	61.9	38.1	45.8	54.2
1951	54.3	45.7	33.5	66.5	29.3	70.7	55.2	44.8	31.9	68.1
1961	50.2	49.8	25.7	74.3	22.7	77.3	42.4	32.4	27.4	72.6
1971	44.1	55.9	19.3	80.7	17.6	82.4	33.0	67.0	24.3	75.7
1981	46.4	53.6	22.4	77.6	18.3	81.7	28.6	71.4	22.0	78.0
1991	49.1	50.9	22.4	77.6	18.2	81.8	25.6	74.4	19.6	80.4

*Includes Nova Scotia, New Brunswick, Prince Edward Island, and Newfoundland
**Includes Manitoba, Saskatchewan, and Alberta
Source: Census of Canada.

The rates of the proportional decline of rural population and the increase of urban population were highest in Ontario, British Columbia, and Quebec. In 1991, Ontario had over four-fifths of its population living in urban centers. Over three-fourths of the populations of British Columbia and Quebec also resided in urban communities. On the other hand, the highest percentages of rural populations were concentrated in the Atlantic and Prairie provinces. Both these regions exhibited a slower rate of proportionate decline of rural populations in comparison with the other regions.

Although it declined rather dramatically as a percentage of the total Canadian population, the rural population did not necessarily decrease in absolute numbers. Indeed, the total rural population of Canada more than doubled between 1871 and 1991 from about three million to over six million people (Table 2.1). But during the same period the urban population grew from less than one million to over 20 million. In other words, the rural population fell far behind its urban counterpart in absolute growth which contributed to its proportional decline.

Anderson (1966), in her study of absolute growth of the rural and urban populations in Canada from 1901 to 1961, showed that the urban growth had been much more rapid in every decade than the rural growth in almost every region. For example, the urban population increased by 61 percent in 1911 over what it was in 1901 and the growth of the rural population during the same period was only 20 percent. While the urban population growth from one decade to the next ranged from 13 to 61 percent, the rural population growth ranged from 8 to 20 percent during the decades between 1901 and 1961. The same trend continued between 1961 and 1971 as well when the urban population grew by 29 percent but the rural population suffered an absolute decline by about seven percent. The trend, however, reversed itself for the first time in the decade between 1971 and 1981 when the rural population growth was higher (14.5 percent) than that of the urban population (12.3 percent). During the decade between 1981 and 1991, however, the urban population again showed a higher rate of increase (13.4 percent) than the rural population (9.8 percent) although all of the increase of the urban population took place in the Prairie provinces and British Columbia; the proportions of the urban populations remained the same in 1991 as they were in the preceding decade in Ontario and Quebec and even went down in the Atlantic provinces (Table 2.2).

Farm and Nonfarm Populations

The farm population was counted by the Canadian census as a separate segment of the rural population for the first time in 1931. The part of the rural population living on census-farms was classified as constituting the "rural farm" population while that not living on census farms was classified as "rural nonfarm." A "census farm" in 1931 was defined as one which was one acre or more in size, farmed directly by a person with or without the assistance of family members or hired labor, and produced at least fifty dollars' worth of agricultural produce in a twelve-month period. The "census farm" was redefined in 1951 to be three acres or more in size regardless of the value of the product, or between one and three acres in size producing at least two hundred and fifty dollars' worth of agricultural produce in a given year. The size of one acre or more again became the basic criterion in 1961 and sales greater than $50, rather than production valued at $50 or more, which was the case prior to 1951, became the economic criterion (Kralt, 1976: 16). A farm had to have sales of agricultural products of $250 or more regardless of the size to be defined as "census farm" in 1981. In 1991, the definition was changed again. The term "census farm" was used for a farm which produced at least one of the following for sale: crops, livestock, animal products, greenhouse or nursery products, mushrooms, sod, honey and maple syrup products. But while prior to 1981 anybody who resided on a census farm was included in the farm population, in 1981 a person had to be a farm operator himself or herself or a member of the family of a farm operator to be included in the farm population. The same definition of the farm population was used in 1991 as well. Obviously, the constraints of definitional change of census farms make direct comparison of the rate of change of farm populations from one decade to another difficult. They, however, do not affect the comparison of trends between census periods as significantly as changes in the definition of rural (Kralt, 1976: 15-7).

The proportion of people living on farms has decreased overwhelmingly between 1931 and 1991 in Canada (Table 2.3), from 31 percent to only three percent. But during the same period the proportion of the nonfarm population has remained more or less the same.

TABLE 2.3

PROPORTIONS OF RURAL FARM AND NONFARM POPULATIONS IN CANADA AND REGIONS, 1931-1991

Populations	1931	1941	1951	1961	1971	1981	1991
Canada							
Farm	31.1	27.1	20.2	11.4	6.6	4.3	3.0
Nonfarm	18.9	21.8	18.2	19.0	17.3	20.0	20.5
Atlantic Provinces							
Farm	40.4	31.4	19.8	8.6	3.8	2.1	1.4
Nonfarm	22.9	31.0	34.5	41.7	40.3	44.3	47.7
Quebec							
Farm	25.9	24.7	18.9	10.7	5.1	2.9	1.7
Nonfarm	15.3	15.4	14.6	15.0	14.3	19.5	20.7
Ontario							
Farm	22.9	18.3	14.7	8.1	4.7	3.2	2.1
Nonfarm	18.3	21.7	14.5	14.5	12.9	15.1	16.1
Prairie Provinces							
Farm	50.4	47.2	37.4	24.0	16.9	11.0	8.6
Nonfarm	19.4	22.2	17.8	18.5	16.1	17.6	17.0
British Columbia							
Farm	14.4	12.3	9.4	4.8	3.4	2.2	1.5
Nonfarm	30.0	34.9	22.5	22.7	20.9	19.9	18.0

Source: Census of Canada.

The proportional decline of the rural farm population has been most significant in the Atlantic and Prairie provinces, the two most rural and agricultural regions of the country. British Columbia and Ontario, which had small proportions of farming populations to start with, exhibited the lowest declines. The decline of farm populations, however, has been sharp in every region between 1931 and 1991 although some regions had a much higher rate of decline than others.

As a percentage of the total population in Canada, the rural nonfarm population is holding more or less on its own. With the exception of the Atlantic provinces, where it showed the most consistent increase, the nonfarm populations in every region either remained stable or showed some, although not significant, decline. As a proportion of the rural population, however, the nonfarm segment showed phenomenal increase not only at the national level, but also at the regional levels (Table 2.4). For example, between 1931 and 1991, the nonfarm population grew from a little over one-third to well over four-fifths of the total rural population. There were similar increases at the regional levels such as in Quebec and Ontario. In the Atlantic provinces, the nonfarm segment of the rural population increased from 36 percent in 1931 to 97 percent in 1991. In the Prairie provinces, where the farm population was proportionately higher than the nonfarm population even in 1971, the nonfarm population grew consistently in proportion to the rural population since 1931 and exceeded the farm population in 1981. On the other hand, in British Columbia, where the nonfarm segment of the rural population has always been higher than the farm segment, the nonfarm population grew consistently from a little over two-thirds of the rural population in 1931 to over nine-tenths in 1991.

The rural nonfarm population, in other words, held more or less on its own as a proportion of the total Canadian population but grew rapidly as a proportion of the rural population in every region since 1931. The rural farm population, on the other hand, declined rapidly both as a proportion of the rural and the total Canadian population. Whatever growth the rural population has shown since 1931, it is mainly due to the growth of its nonfarm segment. Indeed, the rural nonfarm population has actually increased in number from 1,670,116 in 1931 to 5,582,655 in 1991. On the other hand, the rural farm population has decreased from 3,233,400 in 1931 to only 807,325 in 1991. The decline of the rural farm population is thus not only proportional but also absolute. The Canadian rural population today is predominantly nonfarm. The largest decrease of the farm population, however, seems to be over (Harrison and Cloutier, 1995: 26). Between 1981 and 1991 the farm population showed only a 20 percent decrease — the lowest since the 1941-1951 period when it decreased nine percent. Furthermore, the rate of decline was lower (7 percent) for the latter half of the 1981-1991 decade compared to the first half (13 percent).

TABLE 2.4

CHANGES IN RURAL FARM AND NONFARM POPULATIONS
AS A PERCENTAGE OF RURAL POPULATION, 1931-1991

Populations	1931	1941	1951	1961	1971	1981	1991
Canada							
Farm	62.1	55.4	52.5	37.4	27.5	17.6	12.6
Nonfarm	37.9	44.6	47.5	62.6	72.5	82.4	87.4
Atlantic Provinces							
Farm	63.8	50.3	36.4	17.1	8.5	4.5	2.8
Nonfarm	36.2	49.7	63.6	82.9	91.5	95.5	97.2
Quebec							
Farm	62.8	61.7	56.5	41.8	26.2	12.9	7.6
Nonfarm	37.2	38.3	43.5	58.2	73.8	87.1	92.4
Ontario							
Farm	55.5	45.8	50.4	35.8	26.7	17.7	11.5
Nonfarm	44.5	54.2	49.6	64.2	73.3	82.3	88.5
Prairie Provinces							
Farm	72.2	68.0	67.8	56.5	51.3	38.6	33.4
Nonfarm	27.8	32.0	32.2	43.5	48.7	61.4	66.6
British Columbia							
Farm	32.5	26.1	29.6	17.3	13.9	9.9	7.8
Nonfarm	67.5	73.9	70.4	82.7	86.1	90.1	92.2

Source: Census of Canada.

Rural Extinction or Revival?

The rapid proportional decline of the rural population, and the dramatic diminution of the farm population both in relative and absolute terms between 1871 and 1971 led some writers to predict the extinction of a distinctive rural life style in Canada in the near future (Kralt, 1976: 21):

> The rural population, as measured by population concentration and population size, may no longer be as indicative of a distinctive life style compared to a distinctive urban life style. It is very well

> possible, that in fact, the life style of nearly the entire Canadian population is essentially urban in character.

While it is true that the rural-urban differentials in many aspects of life, including demographic, economic, and social, have been narrowing, it is an overstatement to claim that the extinction of a distinctive rural life in Canada is only a matter of time if it has not taken place already. Indeed, the sudden increase of the rural population at a higher rate than the urban population in 1971-1981 emphasizes the hazardous nature of such hasty generalizations.

The rural population of Canada has grown by 19 percent since 1971 — almost as high as the 21 percent growth of the urban population. All regions excepting the Prairie provinces recorded a greater rural population growth. There were almost 500,000 more rural Canadians in 1991 than there were in 1981. Between 1981 and 1991, the rate of growth of the urban population remained the same as was in 1971-1981 in Ontario and Quebec and even declined in the Atlantic provinces. It was only in the western provinces where the urban population showed some growth. The high rate of urban growth appears to have slowed down and the proportional decline of the rural population has stabilized somewhat. As noted above, even the catastrophic rate of decline of the farm population appears to have subsided. There is thus every evidence that a distinctive rural life style, albeit highly modified by urban influences, is here to stay for a long time to come: "Rural Canada endures, is enduring, and will endure" (Bollman and Biggs, 1992: 42).

FACTORS IN RURAL POPULATION GROWTH

The growth of a population depends upon birth, death, and migration, referred to by demographers as "vital processes." The natural increase or decrease of a population is determined by the proportion of births in relation to deaths. A population has a "natural increase" when there is an excess of births over deaths. An excess of deaths over births in a population causes "natural decrease."

Migration as a factor in the growth of a population either reinforces or retards the rate of natural increase. It includes two processes: immigration or inmigration and emigration or outmigration. While the former involves coming to a new place of residence and thus reinforces the natural increase, the latter involves leaving one's place of residence to settle in another, and causes the retardation of the natural increase of a population. The terms "immigration" and "emigration" are used with reference to intersocietal migration while "inmigration" and "outmigration" refer to intrasocietal migration.

Natural Increase of Rural Populations

As a general rule, rural populations have a higher rate of natural increase than urban populations. The high rate of natural increase of rural populations is caused by a much higher rate of birth over death. Most rural societies highly value reproduction of children. Children have always been considered as economic assets by farm families. Many of the important but physically less strenuous chores in farms are performed by children. Mechanization of agriculture in Canada, however, has reduced the importance of children as an economic asset in farming. While most rural societies place a positive value on reproduction of children, urban societies, especially in Western countries, consider small families as desirable.

The 1991 census data show that the rural-urban differential in the fertility rate of women continues to persist. For example, the number of children born to every 1000 married women was 2,203 for the urban population and 2,694 for the rural population. Within the rural population, the fertility rate was higher among the farm women (2,989) than among nonfarm women (2,665). The rural farm population, in other words, had the highest fertility rate, the urban population the lowest, and the fertility rate of the nonfarm population was somewhere between the rural and urban populations. Although the fertility rate has declined for all segments of the Canadian population since 1961, the rural, especially the farm, population continues to have a higher rate than the urban population.

In comparing the rates of natural increase of rural and urban populations during the period between 1921 and 1961, Anderson (1966: 24-5) observed that the rural population in Canada exhibited a higher rate of natural increase (Table 2.5). In the decade between 1931 and 1941, however, the rate of natural increase of the rural population was lower than the preceding decade in almost every region perhaps due to the economic depression of the 1930s. In the following decade all regions showed a much higher rate of natural increase of their rural populations. The highly accelerated rate of natural increase of both the rural and urban populations in the decades between 1941 and 1951, and 1951 and 1961 were the results of the famed post World War II "baby boom."

TABLE 2.5

RATE OF NATURAL INCREASE (PER 1000 POPULATION) IN RURAL AND URBAN POPULATIONS IN CANADA AND REGIONS, 1921-1961

Regions	1921-1931		1931-1941		1941-1951		1951-1961	
	Rural	Urban	Rural	Urban	Rural	Urban	Rural	Urban
Canada	153	126	136	89	173	143	212	181
Atlantic Provinces	118	109	125	104	183	186	220	205
Quebec	206	156	198	115	245	170	246	201
Ontario	114	105	91	68	127	117	204	153
Prairie Provinces	117	144	148	99	169	144	204	205
British Columbia	69	94	61	48	129	107	178	139

Source: Anderson (1966: 24).

Although the urban population also exhibited consistent natural increase during the period between 1921 and 1961 in every region, the rate was much lower than the rural population. Indeed, at the national level the gap between rural and urban populations in the rate of natural increase widened in 1951-61 from what it was in 1921-31. In the more urbanized regions, Ontario and British Columbia, the rural-urban differences in the rate of natural increase grew significantly in 1951-61. On the other hand, in the Prairie provinces, one of the more rural regions, the differential in 1951-61 was almost nonexistent.

Migration of Rural Population

Outmigration from the rural area retards the effect of the higher rate of natural increase of its population than that of the urban area. The movement of people out of rural farms has been a major source of growth of rural nonfarm and urban populations even though they have a lower rate of natural increase. According to Anderson (1966: 25), the farm population suffered a loss in excess of three million people during the decades between 1921 and 1941 through outmigration. Over 60 percent of these people moved to urban centers and 40 percent to rural nonfarm areas.

The magnitude of outmigration from farms varied materially from one decade to another. During the 1920s, there was a net loss of about 472,000 people from rural farms. In the 1930s, when cityward migration from farms slowed down somewhat due to the economic depression, the net loss was 592,000 people. But in the 1940s it rose to 925,000, and reached 1,171,000 in the 1950s (Anderson, 1966: 24-25). The high rate of migration out of rural farms continued as evidence by the consistent proportional decrease of rural farm population in Canada (Table 2.4).

The more rural regions of the country had the lowest rate of net population growth because of a large number of people from these regions migrated to the urbanized regions in search of better economic opportunities. For example, both Ontario and British Columbia, the two more urbanized regions of the country, having a relatively low rate of natural increase from 1921 to 1961, experienced the highest rate of total population growth because of the largest number of people moved to these two provinces from other less urbanized regions.

Quebec suffered one of the highest rates of net outmigration of its rural population even though it had one of the highest rates of natural increase. Quebec, however, maintained a relatively high rate of total population growth as a region since most of its rural migrants moved into its own urban centers, and its urban population also maintained a high rate of natural increase.

The Atlantic region experienced the lowest rate of total population growth because of outmigration of its rural population to urban centers outside of the region.

Its own urban centers attracted few people from the rural areas within or outside of the region. Although the Prairie region also suffered from a heavy outflow of its rural population, the loss was partially offset by a high rate of inmigration to its urban centers (Anderson, 1966: 25).

To summarize, between 1921 and 1961 both Ontario and British Columbia grew rapidly through inmigration from other regions. Their own rural populations migrated mostly into the urban centers located within their boundaries. The urban centers of Quebec and Prairie regions also grew through net inmigrations from rural areas both from inside and outside of their regional boundaries. The Atlantic region attracted very few rural migrants to its urban centers, and lost its rural population through outmigration into urban centers of Ontario, British Columbia and the Prairie provinces.

A more recent analysis (McVey and Kalbach, 1995: 125-8) of interregional migration within Canada between 1981 and 1991 shows the continuation of the same trend: people from more rural provinces and regions moving to those more urban in search for better economic opportunities. British Columbia and Ontario, for example, attracted the highest number of people from other provinces. On the other hand, Newfoundland, Saskatchewan, and New Brunswick — all relatively more rural and agricultural provinces — experienced the heaviest outmigration relative to their populations. While British Columbia, Ontario, and Alberta were the only provinces which gained population as a consequences of interprovincial migration in 1981-91, the remaining provinces either gained very little or even lost population through outmigration of its people to the other provinces.

Reasons for Migration

There are many and varied reasons for migration. Social, cultural, and economic factors of many kinds cause people to move from one place to another. Relatively little research has been done so far in examining the empirical significance of factors believed to be meaningful in terms of relative weights they exercise on migration in different times and in different places. However, ordinarily there is more than one factor which affects the degree to which people respond to an initial stimulus to move.

The so-called "push and pull theory" of migration (Herberle, 1938) attempts to answer the question by classifying the factors which influence people to move into two types: "push" or repulsive factors, and "pull" or attractive factors. The pull factors generally include opportunities for economic and social gain in the place to which an individual migrates driven by the desire to change and improve his or her present social and/or economic situations. The push factors, on the other hand, often include unsatisfactory living conditions in one's place of residence which create a feeling of dissatisfaction in the mind of the potential mover. Economic depression, political disturbances, lack of essential services and facilities, and inadequate employment opportunities are some of the important push or repulsive factors. Mere dissatisfaction with one's present situation, however, is not itself a cause for outmigration. It is usually accompanied by some presumed alternatives. At times, attractive forces give rise to repulsive forces. A person may come to know about better economic opportunities in a different place. When he or she contrasts the available opportunities in the present place of residence with those available in the other place, the person may start to become dissatisfied with his or her home community. In other words, the person may have been quite satisfied with his or her community until he or she learnt of better available opportunities elsewhere.

Anderson (1966: 32), for example, observed that regional differences in economic opportunities were one of the most compelling factors of internal migration in Canada during the period between 1921 and 1961. Population movements between farm and nonfarm areas, and between rural and urban areas within a province were mostly caused by more and better economic opportunities in nonfarm than in farm areas, and in urban than in rural areas. The provinces and regions with relatively high levels of income per capita and a high rate of economic growth attracted people from provinces and regions with a comparatively low rate of income per capita and economic growth. Ontario and British Columbia, for example, showed a high rate of net inmigration of rural people from other regions, for both these regions offered high economic incentives in terms of income per capita and growth in income. On the other hand, the Atlantic region, which was lowest both in income and economic growth, not only failed to attract immigrants from other regions, but also failed to prevent its own rural migrants from moving out of the region.

Movement out of farms, however, can not be solely explained in economic terms. The Prairie region had the highest rate of farm income per capita but it suffered a significant loss of its farm population during 1921-61. Quebec experienced a rapid growth of its farm income but it also suffered a fast decline of rural farm population. The outmigration of the rural farm people in the Prairies and Quebec thus has to be explained in terms of repulsive rather than attractive forces.

One of the significant push factors in farm outmigration is the declining importance of farming as an occupation. The mechanization of farming operations, consolidation of marginal farms into economically more viable units, and the consequent elimination of subsistence farms have pushed a large number of traditional farmers out of farming. In other words, only the more efficient and productive farmers are left on the farms, and the small farm owner, tenant, share-cropper, and surplus agricultural laborer have moved to nonfarm and urban areas in search of a means of livelihood.

Characteristics of Migrants

It is difficult to identify specific traits which characterize people who migrate from one place to another, or are more prone to migrate from rural to urban areas. It is not just that certain types of individuals possessing certain personal traits migrate more often than others. It is possible that push and pull factors are more important than any number of personal characteristics to stimulate the decision to migrate. In other words, individuals with different personal traits may decide to migrate as a response to a number of similar situational stimuli classified as attractive and repulsive forces. Situational rather than personal factors are more important in influencing people to decide to migrate. However, personal factors often act as intervening variables in the relationship between situational forces and the decision to migrate. People with a certain set of personal characteristics are sometimes more prone to be exposed to the situational stimuli which generate decisions to migrate than those who do not possess such characteristics.

A high rate of outmigration from rural, and especially from rural farm areas occurs among young adults. Porter (1961: 140-1) noted that 40 percent of the young people belonging to the age group of 15 to 19 years and living on farms at the time of the 1951 Canadian census left the farms by 1956. A study of population movements in the Atlantic region in 1960 (Levitt, 1960: 30) revealed that the rural and farm outmigrations come largely from the 15-19 age group. In his study of migration among rural residents in Saskatchewan in 1982, Meyer (1992: 311) also noted that the young people in the 18-24 age group showed the highest rate of migration. This tendency of young adults to migrate from rural and farm areas has also been observed in the United States. Bowles (as quoted in Slocum, 1961: 79) offered the following explanation for this tendency:

> Rates are highest for persons just entering working ages for several reasons: 1) they are in search of economic opportunities and have usually formed no or only tenuous job attachments on farms; 2) many have finished high school and will leave farm home to go to college; 3) most are relatively unattached as far as family responsibilities are concerned and thus are free to roam about; 4) their sentimental attachment for farm homes and communities may not be as strong as those of older persons; 5) many of the young people are eagerly in search of new experiences which they feel will be offered in nonfarm areas.

In other words, pull factors operating in terms of availability of profitable employment in nonfarm and urban areas, and push factors operating in terms of lack of attachment to farms lead to high outmigration among young adults. On the other hand, when older people decide to migrate, their decisions are predominantly influenced by push factors operating in rural and farm areas. The mechanization of farm operations, and lack of skill, efficiency, and capital to operate modern farms are the repulsive factors which force older people to migrate. In describing the migration decision process of a group of farmers in Saskatchewan whose average age was 45 years, Abramson (1968: 118) noted that,

> for many, migration occurred not as a positive choice of a preferred way of life, but as a forced choice made necessary by the cumulative deficits of the farm situation. Though all the respondents had, in one sense, migrated voluntarily, about a fifth felt they had no choice. Others felt that they would be forced to leave sooner or later and preferred to move while the choice was still open. Though the economic pressures were of great importance in the decision to leave, social and psychological factors also played a part. A majority stated that the farm situation no longer provided an accepted standard of living and they desired educational opportunities for children, sufficient security or achievement satisfaction, an adequate social milieu, etc.

Studies also show that young female adults tend to outmigrate at a higher rate than their male counterparts. This trend is especially marked in Ontario, the Prairie provinces, and British Columbia where urban employment opportunities for rural young females are many; but the trend is not as significant in Quebec and the Atlantic provinces. In her study of population movement in Atlantic Canada, Levitt (1960: 30) noted that "females tend to migrate at a somewhat higher age than males." Although the reasons are not immediately clear, Whyte (1966: 25) argued that the lack of urban employment opportunities, and high community and family solidarity in the rural areas of Quebec and Atlantic provinces were perhaps responsible for the relatively low rate of outmigration of young female adults from rural and farming areas in these two regions.

A positive association between education and propensity to migrate is often assumed by sociologists. Mobility is increased as members of the labor force acquire superior skills, professional training, and education of all kinds. On the basis of their analysis of the 1991 Census data, McVey and Kalbach (1992: 135) noted that those with less than a Grade 9 education showed a significantly lower rate of interprovincial migration than did those with higher levels of educational attainment. Farm outmigration is similarly accelerated by more schooling in farm areas. Education not only contributes to a higher level of employment flexibility but also to the knowledge of alternative employment opportunities in the mind of an individual (Anderson, 1966: 43). The lack of education, however, may also act as a push factor and affect the decision to move. Indeed, the lack of education and skill needed to operate a modern mechanized farm has forced many farmers out of the farming occupation. Levitt (1960: 30) noted that in the Atlantic provinces "economic and social pressures generate a differentially higher rate of outmigration among groups of people who have lower than average educational attainment." Although she did not have the data to prove it one way or the other, she observed that "within one group of homogeneous economic, social, and ethnic composition, outmigration varies with educational attainment," and that "those people who are more favorably located, who live in urban communities where educational standards are higher and employment opportunities likely to be more plentiful are perhaps under less pressure to move away from home than people located in rural areas where educational standards are lower."

Social and economic factors which generate migration may also act differently on various ethnic and linguistic groups. Stone (1961) provided evidence as to the significance of ethnic and cultural factors such as language and religion in affecting migration rate. He found that during 1951-61, English-speaking Protestants had a higher rate of migration than French-speaking Catholics and Jews. The high rate of outmigration noticeable among certain ethnic and linguistic groups, however, is not necessarily a cultural factor; the fact that these groups are exposed to certain social and economic pressures more than others is enough to generate a high rate of outmigration among them.

In her study of outmigration in the Atlantic provinces, Levitt (1960: 26) found that individuals of French origin outmigrated at a higher rate than those of British origin. Individuals who were French-speaking outmigrated at a higher rate than those who were English-speaking. Among the people with French origin, those who were French-speaking had a higher outmigration rate than those who were English-speaking. Levitt could not find any immediate explanation for such differences, but she emphasized that: "We do not intend to imply that Canadians of French origin have a higher migration rate because they are French Canadians. It is possible that they have a higher migration rate because they are located in areas which for economic reasons produce heavier outmigration."

Adjustment of Migrants

One of the first problems rural migrants confront in a nonfarm or urban area is adjustment to its new environment. The adjustment problem involves not only the change of occupation but also the change of a way of life. It includes establishing contacts with previously unknown individuals in the new area, joining new social systems, and working out adaptations to their customs and social norms.

Abramson's (1968) study of a group of one hundred farmers in the city of Saskatoon provides an excellent example of the adjustment process of rural migrants. The farmers were relatively old with a median age of 45 years. They were forced to leave their farms due mainly to the cumulative deficits of their financial situation.

It took an average of two years for a family to adjust to the urban environment to achieve sufficient economic security and other types of positive attachments. Almost one half of the families did not sell their entire farm after moving to the city, and thus kept the option of returning to the farm open if things did not work out satisfactorily. A significant proportion of the families had previously left their farms for periods of a year or more but went back to their farms.

The major problem of adjustment was occupational and employmental. Unemployment for varying periods and inadequate wages were the major economic problems in addition to the dislike of working conditions in city jobs, dislike of the type of work done, and feelings of lower status and nonachievement. However, such farm related personality traits as the ability to do a variety of jobs, positive attitudes toward physical work, high motivation to achieve economic security, and a stable family situation seemed to have made many of them more successful in adjusting to the urban environment than might have been expected. The group as a whole still appeared to be a disadvantaged one at the time of the study with a disproportionately high rate of unemployment, low income, involvement in unskilled occupations, and lack of any upward occupational mobility.

Four-fifths of the families experienced problems in social adjustment and integration. The majority had problems in adjusting to a system of social relationships characterized by high population density, and formal and impersonal patterns of interaction. The families experienced problems in adjusting with the shifts in work, parental roles, value conflicts between parents and children, and break up of the family as a working unit. Almost all of them sorely missed the large areas of living space and open sky of the farm, of an out-door life, and fresh air and water. They complained about the traffic congestion, and noise of the city life. Sixty percent of the respondents reported psychological stress symptoms caused by the premigration stress, grief of separation from the farm, and the postmigration adjustment.

Based on their level of economic, social, environmental, and personal adjustments, Abramson categorized the farmers into three types: integrated, isolated, and accommodating. The *integrated* type of adjustment was characterized by "a secure economic situation, social participation, positive acceptance of urban environment, and an active, creative personal adjustment." The *isolated* type represented just the opposite extreme in the level of adjustment. "They included those who were economically dependent or unable to provide an adequate standard of living for their families, poorly adjusted to the physical setting or social milieu of the city, and in some cases personally disorganized, unhappy or exhibiting signs of serious psychological stresses." The *accommodating* type constituted the middle group characterized by "at least minimal economic security, a more or less workable balance between their wants, expectations, and aspirations, and the circumstances of their lives, and usually passive affiliative pattern of personal adjustment." Abramson noted that although some improvement in adjustment was likely to occur during the first two years after migration, the level of adjustment did not necessarily improve with increasing length of stay in the city. About a third of the immigrants failed to adjust satisfactorily through a natural process even five or more years after leaving their farms.

There were several premigration characteristics that tended to affect the postmigration adjustment:

> 1) Those who had no previous full-time agricultural jobs were more likely to be found in the Isolated Group. 2) Men who had never been off the farm for more than a few days were more likely to be isolated than those who had previously spent even a few months away... 3) Those whose attitudes toward farm life were negative were more likely to be integrated than those who still had positive attitudes or those who were conflicted. 4) ... Those who had experienced long-term, high-conflict decisions about leaving were much more likely to adjust poorly to the conflict decisions. 5) Those who considered the choice of a new location from many points of view... were more likely to make a good adjustment. 6) A stable, supportive family structure also appears to have facilitated good adjustment during the period of transition. 7) Regardless of level of earnings, those whose postmigration earnings exceeded their premigration expectations were more likely to be in the Integrated Group, while those whose expectations were disappointed were more likely to be in the Isolated Group.

COMPOSITIONAL TRENDS

The rural population of Canada has experienced significant transition in its compositional characteristics over the last several decades. Historically the population composition of rural societies has been different from urban societies, and this difference has been both a cause and consequence of the distinctive patterns of rural-urban life. Although the rural-urban differences in compositional characteristics continue to persist in Canada, the gap has significantly narrowed in recent decades.

Age Composition

Generally speaking, rural societies tend to have a higher proportion of people in the very young and the very old age groups while urban societies tend to include a greater proportion of people in the productive age categories. Rural populations are thus characterized by a higher "dependency ratio" than urban populations because of higher birth rates, outmigration of young adults from rural to urban centers, and the tendency of older people to migrate from city to country after retirement. A small group of economically active people in rural societies carries a rather heavy burden in looking after the welfare of a large dependent population. With the increasing mechanization of farming activities in industrialized countries, the young and the old in rural areas have fewer and fewer economic roles to perform as farming activities become increasingly mechanized.

TABLE 2.6

RURAL AND URBAN POPULATIONS (IN PERCENT)
BY AGE GROUPS IN CANADA, 1971-1991

Populations	Percent in Age Groups								
	0-14			15-64			65+		
	1971	1981	1991	1971	1981	1991	1971	1981	1991
Canada	29.6	22.5	20.9	62.3	67.8	67.5	8.1	9.7	11.6
Rural	33.6	26.5	24.1	58.2	64.7	65.5	8.2	8.8	10.4
Farm	33.3	26.3	25.2	60.8	68.3	67.3	5.8	5.4	7.5
Nonfarm	33.7	26.6	23.9	57.2	63.9	65.3	9.1	9.5	10.9
Urban	28.3	21.2	19.9	63.6	68.8	68.1	8.0	10.0	12.0

Source: Census of Canada

As evident from the 1971, 1981, and 1991 census data (Table 2.6), the rural-urban difference in age composition continues to exist even though the difference is rapidly narrowing. Both the rural farm and nonfarm populations have a larger percentage of very young people (0-14) than the urban, a result of the higher birth rate in rural areas. The urban population has a higher proportion of individuals in the economically productive age group (15-64) primarily because it gains a large number of people into this age group, especially young adults, through rural to urban migration. Historically the nonfarm segment of the rural population has included the largest proportion of people in the old age group (65 and over) because older farm people tend to migrate to rural nonfarm areas after retirement. In recent decades, however, many of these older people from farm areas as well as from metropolitan cities migrated to small towns which have the services and facilities they need. This explains the percentage increase of people in the old age group in the urban population in 1981 and 1991.

The changing age composition is most clearly evident in the dependency ratio (Table 2.7). Although the total dependency ratio (the total number of people in 0-14 and 65+ age groups for every 100 people in 15-64 age group) of the rural population continues to be higher than urban, the difference narrowed considerably between 1971 and 1991. The total dependency ratios of the rural nonfarm were highest in 1971, '81, and '91 followed by those of the rural farm and urban populations. This is the result of the fact that rural nonfarm areas have the highest young (total people in 0-14 age group for every 100 people in 15-64 age group) and old (total people in 65+ age group for every 100 people in 15-64 age group) dependency ratios. Rural nonfarm areas near large towns and cities tend to attract young families with children for their rural environment and inexpensive housing which raises their young dependency ratio. On the other hand, the migration of older people from cities and farms to rural nonfarm areas increases old dependency ratio of the latter.

The comparison of the 1971, '81, and '91 data further reveals that the birth rate has been declining significantly both in rural and urban areas as evident in the sharp fall in the young dependency ratios of all segments of the population. The rural attitudes toward large family size has obviously been affected by the urbanized values. The increase of the old dependency ratios in all segments of the population is a continuation of the trend of increased life expectancy of the Canadian population. The changes in age composition thus can be attributed to the increasing urbanization of the nation. With urbanization, fertility and mortality rates decline; the median age of the population tends to rise for there are fewer young people (0-14), and more productive adults (15-64) and old people (65 and over). Dependency ratios are also affected: the young dependency ratio falls with declining fertility rates while the number of dependent older persons rises as the rate of mortality decreases. The pattern seems to have characterized all segments of the population even though important rural-urban differences continue to persist.

TABLE 2.7

DEPENDENCY RATIOS FOR RURAL AND URBAN POPULATIONS IN CANADA, 1971-1991

Populations	Dependency Ratio								
	Young			Old			Total		
	1971	1981	1991	1971	1981	1991	1971	1981	1991
Canada	47.5	33.2	30.8	13.0	14.3	17.2	60.4	47.5	48.0
Rural	57.8	41.0	36.7	14.1	13.6	15.7	71.9	54.6	52.6
Farm	54.8	38.6	37.4	9.6	7.9	11.1	64.4	46.5	48.5
Nonfarm	59.0	41.6	36.6	16.0	14.8	16.7	75.0	56.4	53.3
Urban	44.5	30.9	29.2	12.6	14.5	17.6	57.1	45.4	46.8

Source: Census of Canada

The Rural Elderly. There have been many studies on the rural elderly in Canada. Keating (1992) provides an excellent review of the findings of several of these studies as summarized below. Although the nonfarm segment of the rural population is significantly larger than the rural farm, "some would still argue that the truly rural elder is the retired farmer" (Keating, 1992: 135). Indeed, the agricultural population in Canada in 1991 was significantly older than the nonagricultural population, and farmers had the highest average age (49.6 years) of all occupational groups (Harrison and Cloutier, 1995: 11). Keating (1992: 136) noted that the labor force participation of farmers aged over 65 was dramatically higher than nonfarm or urban elderly primarily because they were self-employed and perhaps also due to a lack of alternative means of support. Farmers tend to go through several stages before they finally retire from work (Selles and Keating, 1989): the stage of "partial retirement" occurs when the son as a young adult is still at home who becomes involved in the farmwork with his father and is consulted on management decisions; the farmer enters the "semi-retirement" stage after his son's marriage when he transfers a part of his property to his son and allows him to assume management responsibility while he himself assists his son in some farm chores and works with him during the peak agricultural season involving planting and harvesting. As the farmer grows older he reluctantly enters the final stage of retirement because "enforced idleness" due to old age or for other reasons is viewed negatively in rural societies.

Even though the principle of independence is highly emphasized in rural societies, rural elders find it extremely difficult to maintain that independence because of the environment in which they live (Kraut, 1988). Rural elders, especially farmers, live in the oldest housing often built before 1920 (Stone and Fletcher, 1980), and these houses often lack one or several amenities such as running water, indoor toilet, electricity, central heating, and electric stove and refrigerator (Chamberlain, 1976). It has been found that while majority of the rural elderly prefer to and live in their own homes, they vary in the degree of activity and involvement with others; while the "fully" and "partially" engaged rural elders maintain contact with friends and relatives and are involved in community activities, the "disengaged" and "frail" groups are isolated because of social and physical reasons (Scheidt, 1984). There is also variation among the rural elderly in terms of their access to basic community services as bank, grocery store, drug store, doctor, church, social club and so forth. Hodge (1984), who made a list of these community services, found that their availability to rural seniors was greater in small towns with over 2,000 people than in towns and villages with less than 1,000 people. Rural elders themselves, however, perceived and evaluated their immediate environment not in terms of access to services — for many among them, especially farmers, are used to driving long distances for basic services — but in terms of personal attributes such as ability to walk comfortably (Windley and Scheidt, 1988).

Studies conducted on health status of rural elders in Canada consistently show that there is no significant difference between rural and urban elders (Martin Mathews, 1988; Vivian, 1982; Thurston *et al.*, 1982; and Morton *et al.*, 1984). Both rural and urban elders, most of whom have at least one chronic health problem, are, nevertheless, living longer (Keating, 1992: 145). Canadian seniors in general are heavy users of health services although the rate of their use is higher among the urban elderly than their rural counterparts primarily because of lesser availability of such services in rural areas. While in terms of objective measures, the rural elderly appear to be poorly served with health services, "perceived need by rural seniors do not always correspond to objective measures of service availability" (Keating, 1992: 147). They often express low levels of need of health care services and prefer to stay in their homes and communities rather than being moved to an institutional setting in a distant town. The majority of rural seniors perceive that health services available in their communities are adequate or even exceed their needs even though many of their communities are underserved by urban standards or the services are inappropriate for rural needs and culture (Keating, 1992: 148).

Are rural elders part of a supportive family and extended kin group who are close-knit, private, and "take care of their own" including their elders (Cape, 1984)? Based on her analysis of the 1986 census data, Keating (1992: 148) notes that compared to nonfarm and urban seniors, farm elders are most likely to live with their spouses and/or unmarried children and are less likely to live alone. They are also most likely to live in an extended household with people who may or may not be their relatives if they are unable to maintain households of their own. The rural seniors tend to have a larger family network made up of siblings, children, and grandchildren than urban elders which they perceive to be accessible. The degree of their integration to a stable family network depends on whether the rural elderly are long term residents in their communities or are recent immigrants (Wenger, 1986; Harper, 1987).

Older rural couples, as also older Canadian couples in general, are highly satisfied with their marriages as they become increasingly interdependent (Connidis, 1989). This is especially true for farming couples who worked together for the welfare of both the farm and the family throughout their married lives. Keating (1987) found that their husbands' acknowledgment that they have a farm-work as well as a family role is a great source of personal satisfaction to farm wives. Rural seniors as well as their urban counterparts have high levels of trust and regard for their children although they differ in the sources of the quality of that relationship (Keating, 1992: 151). For rural parents the quality of relationship with their children was best if they felt "they had control over their lives; did not turn to the child in an emergency; and did not expect high levels of interaction, provided the child was available." Rural parents who live with their married children in the same household have lower morale and feel that they have a strained relationship with their children perhaps because of the nature of interdependence in such households.

Sex Composition

Until 1971, the "sex ratio" (males per 100 females) in Canada was dominated by males (Table 2.8). While the sex ratio was low (less than 100 males per 100 females) in the urban population, it was significantly high for the rural, especially the rural farm, population. In 1971 the sex ratio achieved almost a perfect balance for the total Canadian population, and in 1981 females outnumbered males, a characteristic feature of an urbanized population. The sex ratio declines with urbanization for several reasons. In traditional societies, high mortality rates among female infants because of parental neglect, and among pregnant women because of lack of facilities for better health care, contribute to high sex ratios. In urban societies both infant female mortality rate and pregnancy-related deaths among women are low because of better health care for pregnant women and equal treatment of both male and female infants. Other factors which contribute to the low sex ratio in urbanized societies are: high rate of mortality among men because of increased occupational hazards and pressures; higher increase in the number of older women than men with the rise of the average life expectancy of the total population because men have a higher mortality rate than women; and the decline of the male-dominated immigrant population as was the case in Canada until 1911 when the sex-ratio reached peak of 113. The decrease in the birth rate is also an important contributing factor to the low sex ratio in urbanized societies because the number of males is greater than females among newborn infants.

TABLE 2.8

SEX RATIOS OF RURAL AND URBAN POPULATIONS
IN CANADA, 1951-1991

Population	1951	1961	1971	1981	1991
Canada	102.4	102.2	100.2	98.0	97.2
Rural	114.0	112.2	109.1	107.1	105.4
Farm	117.4	117.5	115.8	117.0	115.1
Nonfarm	110.3	109.0	106.6	105.1	104.1
Urban	95.8	98.2	97.6	95.7	94.8

Source: Census of Canada

Although women outnumbered men in the total and urban sex ratios in 1991, the sex ratio of the rural population in general and the rural farm in particular continued to be male-dominated. The high sex ratio in the rural farm areas is primarily explained by the difference in employment opportunities farming offers to the two sexes, and by the consequent female-dominant outmigration from these areas. The reason for the moderately high sex ratio in the nonfarm population is more complex. Nonfarm areas adjacent to urban centers lose a large number of females through outmigration to cities while those located near to farming areas attract a female-dominant inmigration of retired people who lived on farms (Norland, 1976: 12-4).

Marital Status

The marital status of the members of a population reflects its demographic, economic, and social characteristics. The rate of marriage in a population tends to increase as economic conditions improve and decrease with economic decline. In Canada, for instance, the marriage rate was high in the 1920s because of good economic conditions and high employment level but it declined significantly in the 1930s due to the economic depression and widespread unemployment. Although the marriage rate rose again with the improving economic conditions after the Second World War, it slowed down in the 1960s because "the relatively small cohorts born during the economic depression came of age which implied a reduction of the supply of marriageable individuals" (Overbreck, 1980: 61). After a rise in the early 1970s as the large number of people born during postwar baby boom matured to marriageable age, the marriage rate started to decline again in the mid 1970s as more and more young people of marriageable age deferred their marriage due to the increased competition in the labor market and tolerance of the society toward alternative living arrangements such as singleness and nonmarital cohabitation (Kalbach and McVey, 1979: 308).

TABLE 2.9

MARITAL STATUS OF RURAL AND URBAN POPULATIONS (15 YEARS AND OVER) IN CANADA, 1981-1991

Population	Married		Single		Separated		Divorced		Widowed	
	1981	1991	1981	1991	1981	1991	1981	1991	1981	1991
Canada	60.9	54.1	27.9	30.7	2.5	2.8	2.7	5.6	6.0	6.5
Rural	65.8	60.8	26.3	27.3	1.5	2.1	1.4	4.4	4.8	5.3
Farm	64.1	67.1	32.4	27.6	0.6	0.9	0.6	2.1	2.3	2.3
Nonfarm	66.2	60.0	25.0	27.3	1.7	2.3	1.6	4.8	5.4	5.7
Urban	59.4	52.0	28.3	31.7	2.8	3.0	3.0	6.4	6.5	6.8

Source: Census of Canada

The trend of declining marriage rate has continued. Little over one-half of the people in Canada who were 15 years of age or over were married in 1991 (Table 2.9).

There was almost seven percent decline in the proportion of married people during the decade between 1981 and 1991. All segments of the Canadian population with the exception of the rural farm population showed significant declines in the proportions of married people during this period. The proportion of married people recorded dramatic fall for the urban population from over two-thirds (67.1 percent) in 1961 to little over one-half (52.0 percent) in 1991. The urban population consistently included a higher percentage of married people than the rural until 1971 when the trend reversed for the first time. A major factor which contributed to the reversal of the trend is the continuing outward movement of families from the more densely populated urban centers into adjacent rural nonfarm areas. Indeed, the rural nonfarm areas had the highest proportion of married people in 1971 and 1981 although the proportion fell by six percent in 1991. The significant increase of the single, separated, and divorced people as well as the increasing trend toward nonmarital cohabitation have also contributed to the decrease of the proportion of married people in the urban population. The proportion of separated and divorced people were lowest in the rural farm areas and highest in urban centers, a reflection of a rather conservative attitude of rural, especially rural farm, people toward separation and divorce and stability in family relationships. Rural farm areas also have the lowest proportion of widowed people because they tend to migrate to nearby nonfarm and urban areas after the death of their spouses. In the past, rural farm populations have had the highest proportion of single people but in 1991 it was the urban population which recorded the highest percentage of the people being single —perhaps a result of a relatively large outmigration of single rural females to urban centers. The rural farm population not only had fewer single people but also was the only population to show an increase in the proportion of married people. Indeed, it had the highest percentage of married people in 1991. Both nonfarm and urban populations showed larger increases than the rural farm population in the proportions of single, separated, divorced, and even widowed people and a consequent decline in the proportions of married people from 1981 to 1991. The rural farm population, on the other hand, showed insignificant, significantly lower, and no increase in the proportions of separated, divorced, and widowed people respectively and a decline in the proportion of single people all of which contributed to the increase of its married people in 1991.

Significant differences in marital status thus continue to persist between rural and urban populations, despite the fact that the rural nonfarm population increasingly resembles the urban population in this characteristic. The marital status of the rural farm population still reflects traditional values even though it has been affected somewhat by the migration of its young female adults and widowed women to urban centers. The urban population has also undergone significant changes in its marital status over the last two decades due mainly to the change in economic situations and social values relating to marriage and the relation between the sexes.

Educational Attainment

Rural populations in general have a lower level of educational attainment than urban populations. The lack of material resources as well as a negative attitude toward formal education in rural societies explain the rural-urban difference in educational attainment. In modern industrial countries, such as Canada, the rate of literacy and the level of formal schooling are high for both rural and urban areas due to increased availability of educational opportunities and resources in almost all parts of the society and a change in the attitude of rural people toward formal education. But, as the 1991 census data show, the rural-urban disparity in educational attainment in Canada continues to persist (Table 2.10).

TABLE 2.10

RURAL AND URBAN POPULATIONS, 15 YEARS AND OVER, BY LEVELS OF EDUCATIONAL ATTAINMENT IN CANADA, 1991

Population	Levels of Educational Attainment (in percent)			
	Less than 9 years*	9-13 years**	Non-University Postsecondary***	Some University or Degree

Canada	13.9	43.0	22.3	20.8
Rural	18.2	47.4	21.2	13.2
Farm	16.9	49.7	20.4	13.0
Nonfarm	18.4	47.0	21.3	13.2
Urban	12.6	41.8	22.7	23.0

*Includes people with no schooling.
**Includes people who earned a trade diploma.
***Requires a high school diploma to attend.
Source: Census of Canada.

The rural-urban disparity, however, is more marked at the two extremes of educational attainment rather than at the middle. The proportion of the urban people with some university education or a university degree was almost twice as large as that of rural people in 1991. The urban population also had a much lower proportion of people with less than nine years of schooling and nine to 13 years of schooling than its rural counterpart. On the other hand, the disparity between the two populations was much less marked in the proportions of people with a nonuniversity postsecondary education. While the occupational composition in rural areas has diversified to the extent that rural people with a secondary education or a trade diploma can find jobs within or near their communities, those with a university degree tend to move to urban centers for better opportunities. The differing economic and occupational environments of the city and the country continue to contribute to the persistence of rural-urban disparity in educational attainment even though the gap has significantly narrowed in recent decades.

The census data of 1981 did not provide the rural farm and nonfarm breakdown of educational attainment. The 1971 data on the rural farm and nonfarm educational attainment showed very little difference between the two populations. Fifty-one percent of nonfarm population compared with about 50 percent of the farm population had a secondary or a higher level of schooling. The 1991 data also show that there is very little difference in the educational attainment of rural farm and nonfarm populations (Table 2.10). Kalbach and McVey (1979: 252) suggest that the presence of the Inuits and native-born Indians in the nonfarm population with only an elementary education or less tend to contribute to the low educational disparity between the farm and nonfarm populations.

Ethnic Composition

Canada continues to be a nation populated predominantly by people of European origin (Table 2.11). The ethnic origin of almost one-half (43.6 percent) of all Canadians were either British or French in 1991. This figure, however, does not include those of British or French descent with multiple origins which would make this proportion significantly larger although perhaps not as large as it was a decade ago in 1981 (67 percent). The 1981 census data, which provided a more detailed breakdown of the ethnic composition, also showed that among the people of "Other European" origin Germans were the largest followed by Italians, Ukrainians, Dutch, Scandinavians, and Polish. The 1991 data show the proportions of the population of German (3.4 percent) and Italian (2.8 percent) descent separately and group the rest of the people of European origin along with those of other descents under "other single origins" (14.7 percent) as well as "multiple origins" (28.9 percent).

TABLE 2.11

RURAL AND URBAN POPULATIONS BY ETHNIC ORIGIN IN CANADA, 1991 (IN PERCENT)

Ethnic Origin	Canada	Rural	Farm	Nonfarm	Urban
British	20.8	24.2	20.3	24.8	19.7
French	22.8	26.4	15.6	27.9	21.7
German	3.4	4.6	10.8	3.7	3.0
Canadian	2.8	2.3	2.8	2.3	3.0
Italian	2.8	0.7	0.4	0.8	3.4
Chinese	2.2	0.2	0.1	0.3	2.8
Aboriginal	1.7	4.3	0.5	4.8	1.0
Other Single Origins	14.7	7.6	14.5	6.6	16.8
Multiple Origins	28.9	29.6	35.0	28.8	28.7

Source: Census of Canada

Minority Canadians, especially those of non-European origin, are largely absent from rural areas and are settled in cities where employment opportunities are better.

The Italians and Chinese, for example, are almost totally absent from rural areas and are found primarily in urban centers. The rural Canadian population is predominantly made up of people of British, French, and Other European origin.

People of German origin constituted the third largest group (10.8 percent) after those of British (24.2 percent) and French (15.6 percent) of the rural farm population in 1991. A decade ago, in 1981, the rural farm population of Other European origin, rather than the French, together constituted the second largest ethnic group after the British. The rural farm population of Other European origin in 1981 was primarily made up of Germans (12.4 percent), Dutch (5.2 percent), Ukrainians (4.5 percent), Scandinavians (2.7 percent), and Polish (1.3 percent). It will be fair to assume that the ethnic make up of the rural farm population in 1991 has not undergone a dramatic transformation from that of 1981 except that most of the Other Europeans have been included in the "Other Single Origins" (14.7 percent) and "Multiple Origins" categories in the 1991 census data.

The rural nonfarm population shows a slightly greater ethnic diversity than the rural farm. The French constitute a larger proportion than the German and even the British. The aboriginal population of Canada belongs predominantly to the rural nonfarm population. The 1981 census data showed that the rural nonfarm population also included small proportions of people of Asian, South American, and African descent.

To summarize, the urban population is ethnically the most heterogenous of all even though it is dominated by people of European descent. Non-European ethnic groups are mostly concentrated in urban centers. The rural nonfarm population falls in the middle between the rural farm and urban populations in ethnic heterogeneity.

Occupational Composition

The terms "rural" and "agricultural" are no longer synonymous in Canada. Only 15 percent of the total rural population in 1981 followed agriculture as an occupation. A little over one-half of the people who lived in farm areas were agriculturists by occupation. The remaining one-half of the "farm" people followed a wide variety of nonfarm occupations.

The nonfarm population was almost totally nonagricultural in 1981 with about 90 percent working in secondary and tertiary industries. Close to two-thirds of the total rural nonfarm people worked in manufacturing, processing, service, and retail industries. With the exception of a little over one-tenth of the people who worked in primary industries, the rural nonfarm segment of the Canadian population was very similar to the urban population in occupational composition in 1981.

The picture remained essentially the same in 1991 (Table 2.12). About one-fifth of the rural population was employed primarily in agriculture in addition to other primary industries — fishing, mining, and logging. Over one-half (58.8 percent) of the people belonging to the farm population worked in the primary industries including agriculture while the remainder followed nonfarm occupations.

TABLE 2.12

RURAL AND URBAN POPULATIONS BY OCCUPATIONS IN CANADA, 1991

Occupation	Canada	Rural	Farm	Nonfarm	Urban
			(in percent)		
Primary industries	6.1	17.9	59.8	9.5	2.7
Manufacturing industries	14.7	13.8	5.0	15.6	14.9
Construction industries	6.7	8.2	3.6	9.1	6.1
Transportation storage industries	4.1	4.6	2.4	5.1	3.9
Communication and other utility industries	3.4	2.6	1.2	3.0	3.6
Trade industries	17.2	14.9	7.2	16.5	17.9
Finance, insurance, and real estate industries	5.7	3.4	1.9	3.6	6.4
Government service industries	7.8	7.1	3.2	8.0	8.0
Educational service industries	6.8	6.2	4.2	6.6	7.0
Health and social services industries	9.0	7.5	5.1	8.0	9.4
Other industries	18.7	13.7	6.5	15.1	20.1

Source: Census of Canada.

Nine out of ten people of the nonfarm population were employed in industries other than primary. With the exception of the fact that about 10 percent of the people in nonfarm areas worked in primary industries, the nonfarm population closely resembled the urban population in occupational pursuits in 1991 (Table 2.12). The proportions working in manufacturing, communication and other utility, trade, and various service industries are very similar for the nonfarm and urban populations. Since the vast majority of the rural people belong to the nonfarm segment (87 percent), it has contributed to an increasing convergence of rural and urban populations in occupational composition. This convergence is largely a function of the technological revolution in agriculture, and establishment of manufacturing, processing, and service industries in rural areas. While a large number of farmers, squeezed out of farming, moved to urban centers for employment, those who stayed behind took employment in the newly developed industries in rural areas. The occupational shift, in turn, contributed to significant changes in the rural social structure of Canada.

SUMMARY

The rural population in Canada has undergone significant changes both in distribution and composition. Although it suffered a continuous proportional decline during the 100 years between 1871 and 1971, the rural population showed a proportional increase for the first time in Canada's history in 1981. While it decreased again in proportional terms in 1991, the rate of decline was low and evident only in the western provinces. The rural farm population, however, continued to decrease both in absolute and relative terms in 1991 although it showed the lowest rate of decline since 1951. The mechanization of farming and technological transformation of agriculture make fewer and fewer people necessary to produce food and fibre in an industrialized society. Small and subsistent farmers who failed to adapt to this transformation have been forced out of farming, and in many cases from their rural communities, many of whom migrated to urban centers searching for an alternative means of livelihood. The move from farm to nonfarm occupation, and from a rural to an urban environment has neither been easy nor very happy for many of these rural migrants.

The urban influences and the movement of people from the rural to urban areas have altered the traditional composition of the rural population. The young dependency ratio has declined with the decreasing fertility rate of women. The old dependency ratio also shows a declining trend as more and more older people, especially widowed women, move from the rural to urban areas in search of better services and facilities. The national trend of decreasing sex ratio is reflected in the rural sex ratio although men still continue to outnumber women in the rural areas. A major reason for the higher sex ratio in the rural areas is the tendency on the part of young female adults to migrate to cities in a higher proportion than males in search of employment.

The low sex ratio in the urban areas and high sex ratio in the rural areas respectively account for a higher proportion of married men in cities and of married women in the country. Although the proportions of single, separated, and divorced people have increased markedly in the urban areas due primarily to increased societal tolerance to singleness, divorce, and nonmarital cohabitation, such increases have been less marked in the rural, especially farm, areas reflecting a more conservative orientation of the rural people.

The rural-urban difference in educational attainment has narrowed rapidly in recent decades with the increasing availability of material resources and facilities for a better schooling and a change in rural attitudes toward formal education. The disparity, however, continues to persist in the lowest and highest extremes of educational attainment because of the continuing rural-urban differences in economic and occupational environments.

Rural Canada is still ethnically a homogeneous society with the people of European descent almost entirely constituting its total population. Urban Canada, on the other hand, shows a higher degree of ethnic heterogeneity as new immigrants, especially those of non-European origin, mostly settle in cities where employment opportunities are better than rural areas.

Perhaps the most important change to have taken place in rural Canada in recent decades is the transformation of agriculture as an occupation. The agriculturists constitute a small minority in the rural population, and only a little over one-half of the farm people practice farming as a means of livelihood. Although many important differences persist, there has been an increasing convergence between rural and urban Canada in occupational composition. The technological transformation of agriculture and the change of its role as an occupation of the majority of the rural people are indeed two of the most important factors which have had far reaching consequences both for the rural economy and rural social structure.

CHAPTER THREE

Changing Agriculture: Contributing Factors

Although much fewer in number, Canadian farmers are more efficient and productive than ever before. Several factors have contributed to the rise in their efficiency. Development and use of new farm practices have revolutionized agricultural technology. The modern agricultural technology has drastically altered the pattern of farm input: tools and machines are increasingly being used as a substitute for labor, and there has been a phenomenal increase in the application of farm chemicals such as fertilizers and insecticides. With the change in the pattern of farm input has come a new orientation and attitude toward farming on the part of Canadian farmers. Farming has become a "way of making a living" with primary emphasis on economic and commercial considerations rather than a "way of life," a characteristic of peasant agriculture in which social, familial, and religious considerations are also of equal importance.

MODERNIZATION OF CANADIAN AGRICULTURE

Increased productivity is a major indicator of agricultural modernization. Agricultural productivity has been growing steadily inspite of the dramatic decline in the number of farms and farmers in recent decades.

Agricultural Productivity

Yield trends in crops and livestocks provide a straightforward measure of the growth in agricultural productivity. Daviult (1980: 55) noted that over almost a forty-year period between 1935-39 and 1978, per acre yields of wheat, potatoes, and mustard seed grew more than two and a half times; yields in corn, barley, and oats doubled; that of rye tripled; and the increase in the per acre yield of flax seed had been fourfold. Based on his study of the yields of wheat and barley, Veeman (1995: 519) also found that between 1920 and 1993 wheat yield grew by more than two times and the increase in the yield of barley was even higher. Rapeseed production showed a tenfold increase between 1967 and 1997.

Similar increases have been recorded in livestock and poultry. For example, although the number of dairy cows in Canada dropped by 32 percent, yield of milk per cow rose 73 percent between 1950 and 1975. During the same period, the number of beef cattle in Canadian farms more than tripled; hog production increased by over 50 percent; and poultry meat production showed a threefold growth (Task Force on Canadian Agriculture, 1977, v.1, part A: 115). More recent data confirm the continuation of the same trend of increased productivity. Between 1971 and 1996, for example, the number of dairy cows dropped by almost 50 percent but milk production per cow increased by almost 66 percent (Boyd, 1999: 71). Hog production increased by 80 percent and pork production 23 percent during the decade between 1988 and 1998, even though there was a 45 percent drop in the number of farms producing pigs (MacGregor, 1999: 107). Similar trend of fewer farms producing more and more is discernible in the production of beef and poultry as well (Plourde, 1999: 99-105; Negrave, 1999: 93-7).

A widely used and a more accurate measure of agricultural productivity is "input-output ratio." An increase in productivity is indicated if output exceeds per unit of input. Farm "inputs" include labor, real estate, and such capital inputs as machinery, livestock, feed and seed, fertilizers, and pesticides. "Output" is the value and volume of products grown for sale in market. The increase in total output in Canadian agriculture has far exceeded the increase in total inputs over the last several decades.

A study of agricultural productivity from 1935-1939 to 1960 (Furniss, 1964) has shown while the total farm output grew by 49 percent, the total farm inputs rose only seven percent during this twenty-five year period. The average annual growth of agricultural productivity was 2.6 percent, and the overall increase was 60 percent higher during this period than that of the preceding decade. The same trend was observed in another study conducted for the period between 1950 and 1969 (Furniss, 1970). The rate of growth of the total farm output continued to be much higher than that of the total farm inputs. Between 1960 and 1969, the average annual rate of growth in agricultural productivity was 2.5 percent. A more recent review (Veeman, 1995: 52) of annual growth rates during the 44-year period between 1948 and 1991 also showed significantly higher rate of output compared to that of input and a high rate of growth of agricultural productivity each year: "aggregate output rose by 1.86% per annum in Canadian agriculture, aggregate input use declined marginally by -0.02% per year, and productivity increased by 1.88% per year over the 44-year period." In other words, the overall productivity of Canadian agriculture, measured in terms of input-output ratio, has increased phenomenally over the last several decades.

Productive Efficiency of Farmers

The increasing efficiency of Canadian farmers played a significant role in the growth of agricultural productivity. The dramatic growth in agricultural labor productivity is evident in the fact that between 1949 and 1978 the output per person employed in agriculture increased four and a half times. A study (Daviult, 1980: 138) conducted in 1978 showed that since the early 1960s, labor productivity in agriculture, measured both in terms of the value of output per dollar of labor input, and output per man-hour, consistently exceeded those in other sectors of the economy. In 1978, for example, the index of output per person was higher in agriculture (126.7) than nonagricultural goods producing (118.7), commercial services producing (109.2), and manufacturing (125.9) sectors. As a consequence of the dramatic growth of the labor productivity in agriculture, fewer farmers are producing more food than ever before: "In 1935, an average Canadian farm produced enough food for 11 people; by 1994, enough for 123 people" (Bristow-Callahan, 1999: 23).

The increase in farming efficiency and productivity, however, is not universal among Canadian farmers. According to a 1980 study (Brickman, 1980), only about 30 percent of all Canadian farmers, who are large operators, produce 80 percent of the total production. Another 30 percent are primarily "hobby farmers" who produce about two percent of all gross sales. The remaining 18 percent is produced by "limited resource" farmers (40 percent of all farmers) who operate small farms with little capital and are often traditional in skill and orientation. It is thus the large commercial farmers who are mainly responsible for modernization and increased productivity in agriculture. The majority (70 percent) operate relatively small holdings, produce only 20 percent of the total production, and depend primarily on off-farm income.

FACTORS IN MODERNIZATION

Three factors have contributed to the increased agricultural productivity in Canada (Auer, 1975): (1) movement of labor out of agriculture; (2) increased capital and material input; and (3) a host of other indirect and more or less quantifiable factors such as increased farm size, commercialization and specialization of farming, changing marketing structure, increased knowledge and education of individual farmers, and research and development of new farm practices.

Decline in Labor Input

In his study of agricultural productivity in Canada, Auer (1975: 24) noted that the most significant contribution to growth in labor productivity came through adjustments in agricultural employment. With the retirement of older farmers and movement of rural youths to nonagricultural employment, some of the low-productive marginal farms have either been abandoned or have reverted to forest land, and some of the better farms have been absorbed into larger farm units. The increasing use of machines and equipment make operation of these enlarged farm units possible without additional farm labor. Thus there has been a gain in labor productivity due to the outmigration of farm labor and mechanization of agriculture.

Labor input declined drastically over the last several decades. Labor constituted about two-thirds of the total input in agriculture in the mid-Thirties but decreased to only one-third in the early Sixties and to less than one-fourth in the mid-Seventies (Shute, 1975: 4). The substantial increase in the relative cost of labor since 1940 acted as an incentive to Canadian farmers to substitute machinery for labor as a cost reducing device. Of the three categories of farm labor, self-employed operators, unpaid family workers, and paid workers, the first two declined both in absolute and relative terms since the mid-Sixties.

Increase in Capital Input

Capitalization as a process involves substitution of inputs such as machinery, fertilizer, pesticides and feed and seed for labor. Such a shift means increasing substitution of money for labor.

Mechanization

The use of mechanical power perhaps had the greatest impact on Canadian agriculture. Mechanization not only revolutionized agricultural productivity but also affected rural life in Canada in a significant way.

In providing a historical sketch of the evolution of use of machine power in agriculture, Abell (1968) noted that until 1830 farming methods in Canada were primitive. Axes, spades, and hoes were the main agricultural tools which were not only inefficient but also wasteful in terms of human labor. In 1830, the old Loyalist ploughs made almost entirely of wood were replaced by American cast iron and Scottish wheelless ploughs. Lighter and more efficient harrows came to be widely used in the 1840s instead of the traditional harrows which were heavy and unwieldy. Steam tractors, threshing machines, and multiple hitch combines came in during the later part of the nineteenth century.

The mechanical revolution in farming was stimulated further by the First World War as farmers spent more and more money they earned from higher prices for farm products in buying farm machinery. Three mechanical innovations which came about in the 1930s and 1940s had the greatest impact on Canadian agriculture. Putting rubber tires on farm machines lifted farm operations above the walking gait of man and horse. Fitting implements to power units instead of pulling them by tractors contributed to more efficient use of implements. The third innovation was mounting of power units on machines as is the case with self-propelled combines, lettuce pickers or fruit pruning and picking machines.

More recent developments involve still more efficient machines. On the dairy farm, for example, milking machines are connected to cooling tanks from which milk is picked up by a refrigerated tank truck to be delivered to a milk processing plant. The whole operation is conducted without being handled physically by humans. Poultry raising has been mechanized to the point that a single farmer can raise thousands of birds.

The use of such mechanical labor saving devices is possible due to the increasing availability of electrical power. Extensive electrification of farms is a relatively recent development in Canada. Even as late as 1951, only one half of all farms in Canada had electric power. However, by the late 1970s over 90 percent of the farms had electricity. Although the availability of electricity on a farm does not necessarily mean that it will be used extensively for farming purposes, the increase in availability of electrical power has been accompanied by a rise in the use of electric motors in Canadian farms. In 1941, there were only eight electric motors per 100 farms in Canada. By 1981, well over 80 percent of the farms used an electric motor.

The rise in the use of farm machinery has been phenomenal (Table 3.1). Between 1941 and 1991, the number of motor trucks increased almost seven times; the number of tractors more than quadrupled; and the number of grain combines increased by eight times even though their number was almost nine thousand fewer in 1991 than what it was in 1981. The numbers of pick-up hay baylers and forage crop harvesters doubled since they were first introduced in 1961. As was the case with grain combines, the number of forage crop harvesters in 1991 dropped by almost one thousand from that in 1981. Replacement of smaller and less efficient machines by larger and more efficient ones appeared to have contributed to the decline in the numbers in both instances.

TABLE 3.1

NUMBER OF SELECTED MACHINES ON FARMS IN CANADA, 1941-1991

Year	Motor Truck	Tractor	Grain Combine	Pick-up Hay Bayler*	Forage Crop Harvester*
1941	77,480	159,752	19,013	--	--
1951	196,122	399,686	90,500	--	--
1961	302,013	549,789	155,611	89,522	16,764
1971	369,849	596,698	162,751	152,832	28,534
1981	474,408	657,606	161,110	165,341	35,756
1991	526,808	734,149	152,114	167,126	34,654

* Pick-up hay bayler and forage crop harvester came into use in 1961.
Source: Census of Canada

The number of farms using computers has grown dramatically since 1986. Computers are used most frequently by the largest farms in Canada with a gross farm income of $250,000 and over. The proportion of these farms using computers grew from little over one-tenth in 1986 to almost one-half in 1996 (Lonmo, 1999: 239-44). The proportion of all farms using computers increased from only two percent to 21 percent during the same ten-year period. Computers are used not only for record keeping and accounting but increasingly also for weather forecasts, commodity price information, discussion on farming issues on "chat" sites, and specialized information on particular types of farming.

Fertilizers and Chemicals

Almost 50 percent of the total increase in crop production is attributable to the increased use of chemical fertilizers. There has been a fourfold rise in the overall consumption of fertilizers by Canadian farmers since 1960. The consumption of pesticides has more than quadrupled during the post World War II period. Increased acreage in crops and growing size of the livestock herd are the two major contributing factors to the growth of consumption of farm chemicals. The control of insects, plant diseases, and weeds with the help of improved varieties of insecticides, fungicides, and herbicides had a very important role to play in the rise of agricultural productivity.

Seed and Feed Input

The use of pedigreed and disease-resistant varieties of seed have contributed to high yields in crop production. For example, the dramatic rise in corn acreage and yield in Canada since 1950 can be attributed to the introduction of new high yielding hybrid varieties which mature under cooler and shorter growing seasons. Specialized techniques developed in seed production through scientific means has made farmers rely more on the commercial seed producing industry rather than their own seeds.

With the increasing size of livestock and poultry operations, the expenditure on commercial feeds has grown phenomenally. Canadian farmers annually spend over a billion dollars on commercial feeds which represent about one-fifth of their total farming expenses. The sale of commercial feed had a ten-fold increase since 1950 (Task Force on Canadian Agriculture, 1977, v.1, part A: 145).

The rising agricultural productivity and farming efficiency thus can be traced directly to what is called the "changing input mix." Agricultural productivity involves the relationship between what is produced (output) and what is used (input) to obtain that production. If output can be increased without raising input, or without increasing input in the same proportion, then growth in productivity is achieved. A major factor in the changing input mix is reduction of labor and increase in capital investment. The increased use of machinery, fertilizers and farm chemicals, and improved qualities of seed and feed have contributed significantly to the rise in input-output ratio and productivity of agriculture.

Other Factors

There are other factors which are not directly related to the expenditures on the farm and cannot be readily measured, but they contributed to the rise of agricultural productivity almost as much as all material and capital input combined (Auer, 1975: 25). These factors are: increased farm size and scale of operation, commercialization and specialization of farming, changing marketing structure, increased knowledge and education of farmers, and research and development of new farm practices.

Farm Size

The average size of farms in Canada has increased two and a half times since 1941 (Table 3.2). The same basic pattern is observed in all regions of Canada except in the Atlantic provinces and British Columbia where the average size of farms declined slightly between 1971 and 1981 but increased again in the decade between 1981 and 1991. The Prairie provinces continue to have farms much larger in size than the national average because of the extensive and commercial nature of their wheat farming.

The proportion of large farms has increased with the rise in average farm size. While 64 percent of all farms in 1951 were below 240 acres in size, in 1991 this proportion fell below 50 percent. Thus over one-half of all farms in 1991 were 240 acres or more in size. Most of these large farms came into existence through horizontal integration of smaller and less efficient units with larger and more efficient farms rather than through inclusion of additional land which was not previously cultivated. Indeed, the total area under farming in Canada showed almost ten percent decline in forty years: from 173,924,000 acres in 1941 to 162,815,073 acres in 1981, although it increased to approximately 167,419,000 acres in 1991 and 168,164,000 acres in 1996.

TABLE 3.2

AVERAGE SIZE OF FARMS (IN ACRES) IN CANADA
BY REGIONS, 1941-1991

Regions	1941	1951	1961	1971	1981	1991
Canada	237	279	359	463	511	597
Atlantic Provinces	116	135	185	244	234	258
Quebec	117	125	148	176	194	222
Ontario	126	139	153	168	181	196
Prairie Provinces	405	498	617	765	841	952
British Columbia	153	178	226	316	269	307

Source: Census of Canada

With the increase in average farm size, there has been a dramatic decline in the number of farms since 1941. The number of occupied farms fell from 732,832 in 1941 to 280,043 in 1991 and 276,548 in 1996. In other words, 456,284 farms went out of existence during this fifty-five year period between 1941 and 1996 at the rate of 8,296 farms per year, an overwhelming testimonial to the impact of advancing technology. Increase in farm size and reduction in the number of small farms of subsistent type contributed significantly to productive efficiency and growth in agriculture.

Commercialization and Specialization

One of the most important changes associated with modernization of agriculture is the highly commercialized nature of farming. Commercial farming is far removed from the traditional subsistent type of activity. Subsistent farmers produce barely enough to satisfy their own consumption needs and sell the little surplus they have left for cash to buy items they do not produce. Commercial farmers, on the other hand, produce exclusively for the market and profit. The major concern of commercial farmers is to maximize return on their labor and capital investments. Substitution of labor with capital, mechanization of farming activity, and increase in farm size and scale of farming operation, are related to this increasing commercial and market oriented farming enterprise.

The commercial nature of Canadian farming can be measured in terms of value of products sold. The proportion of farms producing more than $2,500 worth of goods for sale in a year, defined as "commercial farm" for census purposes, has grown significantly just during the two decades between 1971 and 1991 (Table 3.3). While the value of products sold was less than $2,500 for almost 30 percent of all farms in 1971, the proportion was only 8.5 percent in 1991. A more dramatic measure of commercialization is the fact that 42.3 percent of all Canadian farms produced at least $50,000 worth of goods in 1991 while the corresponding proportion in 1971 was only 2.9 percent.

TABLE 3.3

FARMS CLASSIFIED BY VALUE OF PRODUCTS SOLD IN CANADA, 1971-1991

Value of Product	Percent of Farms			
	1971	1976	1981	1991
Under $2,500	29.3	20.6	14.7	8.5
Over $2,500	70.7	79.4	85.3	91.5
$50,000 and over	2.9	12.9	28.5	42.3

Source: Census of Canada

Although the trend of increased commercialization is observed in all regions of Canada, it is most extensive in the Prairie provinces and in Ontario, the former being characterized by extensive wheat farming on a large scale and the latter by small scale specialized farming of an intensive nature. The rate of commercialization is relatively low in Quebec and the Atlantic provinces primarily because of the predominance of small scale subsistence farming in these two regions. British Columbia also had a relatively low rate of increase in commercial farming. To a large proportion of farmers in this region farming is a part-time activity, especially in the areas outlying the urban centers.

Farming in Canada is also becoming a specialized activity. The traditional mixed farming is almost a thing of the past. The average Canadian farmer is no longer a subsistent generalist; he or she is a specialist in a few farm enterprises. Several factors contributed to specialization in farming (Abell, 1968: 203). The need for economic use of highly specialized farm machinery has compelled farmers to specialize in their farming activities. With the greater availability and use of scientific knowledge about chemical content of soil, fertilizers to use to compensate for soil deficiencies, and climate requirements for crops, farmers no longer need to diversify their farming operations for security against uncertainties. Growing internal and external markets have also provided incentives to farmers to specialize their productive efforts. Finally, the increasing productive efficiency and profitability in farming has been a major motivational factor for specialization.

Marketing Structure

Changes in the marketing structure have been an important part of the process of farm modernization. A century ago, the Canadian farmer was his own man: he not only produced agricultural goods on the farm he owned and operated with the help of family labor and hired hands, but also processed, transported, and marketed his produce himself. Most of these tasks have become highly differentiated and specialized today. The Canadian farmers now grow a great volume of food and fibre in their large, specialized, mechanized, and commercialized farms for the ever expanding market over a great distance. The modernized marketing structure helps them to assemble and transport their products to national and international markets. The marketing channels for primary agricultural products can be discussed in following terms (Task Force on Canadian Agriculture, 1977, v.1, part A: 167-93):

(1) Producer Sales; (2) Processors; (3) Wholesalers; (4) Retailers; and (5) Industrial and Export Markets.

Producer Sales. Producer sales include direct sale by farmers to buyers; public exchange; collective marketing; and marketing through middle men. Direct sales are of three types: farm to farm, farm to consumer, and farm to processor. Farm to farm sales primarily include sale of seed and feed grains as well as pigs and calves for breeding. Farm to consumer sales involve direct sale of goods to consumers by producers such as in roadside stands and farmers' markets. Direct farmer to processor sales often takes the form of "contract farming," which is one of several types of "vertical integration." Contract farming involves an agreement between a farmer and a processor, wholesaler, or retailer of farm products to partially coordinate the supplying, producing, processing, and marketing of a farm product under one management.

Public exchanges provide facilities, hours of trade, and terms for sale of agricultural products. Terminal markets held in major cities are a type of public exchange where agricultural goods are sold by commission agents on behalf of farmers to processors, wholesalers or retailers primarily through auction or negotiation. Auction markets are held locally, and owned and operated by private interests, who sell directly to buyers on behalf of farmers.

Cooperatives are becoming increasingly important for collective marketing. Farmers' cooperatives are often involved not only in wholesaling, retailing and processing activities but also in production and distribution of farm inputs in competition with private firms. Another kind of collective marketing is exemplified by marketing boards at provincial levels which are often producer-controlled. The directors of these boards are elected by farmers who administer marketing plans and regulate production, distribution and marketing prices.

Sale through merchant, speculative, or agent middlemen is another marketing mechanism. Both merchant and speculative middlemen take title to the goods they handle. Private owners of grain elevators are an example of merchant middlemen who buy grains from farmers, clean, grade, and store them to sell to local farmers or to ship to other regions or export markets. Speculative middlemen do little or no processing or cleaning of the merchandise they buy; their objective is to make profits through short-term price changes. Agent middlemen, on the other hand, do not take title to the goods they handle. They receive fees and commissions for providing market information and getting the buyers and sellers together.

Processors. Food processing is the largest among the secondary industries in Canada both in terms of employment and production. Food processing plants, located close to producing areas to minimize transportation costs, buy agricultural produce either directly from farmers in terminal and auction markets or they themselves get involved in the production of raw materials which they process. Although processing companies are largely privately owned, farmers' cooperatives also operate processing plants.

Wholesalers. The wholesaling function involves assembling and/or further manufacturing of products for sale to retailers. The wholesaling function, however, is increasingly being taken over by processors and retailers. In poultry and dairy industries, for example, processors often buy directly from producers the raw material which they process before selling them to retailers. Large retailers are also taking over wholesaling functions as they perform much of the assembling function by dealing directly with producers.

Retailers. Retailers provide the consumers with food products in small quantities. Their commodities range from slightly altered to highly processed meals served to restaurants and hospitals. Food retailing in Canada is being increasingly concentrated in the hands of a few large firms. One major problem involved in this kind of concentration is that they often assume the power of dictating prices of commodities both to farmers and consumers. On the other hand, the increasing influence of retailers in food marketing keep the powerful processors under control.

Retailing of food commodities is mainly done by retail chain stores, and hotels and restaurants. Over one-fourth of the total retail sales in Canada is accounted for by the retail food sector. As the overhead expenses grow, larger multi-service retail outlets replace smaller corner grocery stores. Almost two-thirds of grocery sales in Canada are done by four major corporate organizations. Large hotel and restaurant chains and franchises are also growing in number. The top 50 hotel and restaurant firms in Canada account for about one-half of total sales.

Industrial and Export Markets. Agricultural products are used for industrial purposes as well. Tobacco and fibre, for example, are major agricultural products in this category. Large volumes of agricultural products are also being exported for international markets. Canada is the second highest exporter of wheat in the world. Canadian wheat accounts for 20 percent of the world's wheat exports. Barley, rapeseed, and flaxseed account for about 20, 60, and 80 percent of world exports respectively.

Export of agricultural products is increasingly becoming a very important contributor to the growing Canadian economy. In 1997, exports of agricultural products accounted for seven percent of all Canadian exports. Canada's agricultural and food trade accounted for three percent of the agricultural and food trade in the world in 1997 and is expected to create up to half a million jobs by the year 2005. The main commodities exported by Canada in 1997 were grain and meat products and rapeseed which together accounted for more than one-half of the total agricultural exports. In 1997, "over one-third of farm-gate receipts were coming from exported products. One of the consequences of this success, however, is that Canadian farmers have become increasingly dependent on international markets for their economic well-being" (Basillais, 1999: 50).

Significant improvements in marketing technology have also contributed to a reduction in marketing costs. The steel rolling mill, the box car, and the vertical grain elevator, for example, have reduced the cost of marketing livestock. Truck transportation, refrigeration, prepacking, self-service distribution outlets, bulk milk handling, and dehydration and freezing of vegetables and foods have contributed to the improved marketing of farm products in Canada (Wood, 1962). All these marketing activities, however, are increasingly carried out off the farm and thus have reduced farmers' share of the consumers' dollar. Indeed, about 65 cents of every dollar spent by consumers go toward the marketing cost.

Vertical Integration

Since vertical integration is increasing within the agricultural marketing system and its effect on Canadian agriculture is controversial, it will be appropriate here to discuss its nature and consequences in some detail. Although vertical integration has helped in accelerating the farm modernization process, its effects have not always been beneficial to the interests of farmers.

There are two general types of integration: horizontal and vertical. Horizontal integration in agriculture is simply the merger of two or more farms into a larger unit. It has been noted that the increase in average farm size has taken place through the process of horizontal integration rather than through addition of previously uncultivated land to farming.

Vertical integration, on the other hand, is the combining of two or more successive stages in production and/or distribution under the ownership or control of a single management. There are two general types of vertical integration: ownership and nonownership.

One type of ownership vertical integration (also called "forward" integration) is achieved through collective action by farmers themselves in which members of a farmers' cooperative coordinate the production, processing, distribution, and even retailing of their products. This enables farmers to retain independence and achieve economic security at the same time.

The second type of ownership vertical integration (or "backward" integration) results when a processing or a retailing firm owns or leases the factors of production (such as land, equipment and other inputs) and hires its own employees (who are often formerly independent farm operators) to tend the required livestock or crops. Although backward integration enables these firms to control the quantity and quality of production needed to maintain an efficient operation, it presents a threat to the autonomy and often survival of relatively small and independent farmers.

Nonownership vertical integration can also be classified into two types. The first type exists when the governments take over part of the farm management decisions by fixing prices and legislating marketing procedures. Provincial governments, for example, control marketing boards through legislative means. The federal government of Canada also plays a very important role in the marketing of agricultural products which primarily centers around quality standards, price of certain agricultural products, trade and transportation policy, farm incomes, and consumer welfare.

Contract farming is the second form of nonownership integration. It is an agreement between a farmer and a firm to partly coordinate the supply of inputs, production, processing, and marketing of a farm product under one management. There are several variations of contract farming. A vegetable farmer may enter into a contract with a processing firm to raise a certain product with a specified quality of seed and fertilizer which he or she might be obligated to buy from the firm and sell the product to it at a previously agreed upon price when the crop is harvested. In another type of contract farming, the firm extends credit to the farmer but no guaranteed price for the finished product. In a third type, found especially in the broiler industry, the company provides most of the operating resources, owns the commodity being produced by the farmer, and assumes most of the managerial functions of production. The farmer primarily contributes labor in the production process and is paid on the basis of output. Although a contractual agreement of this type limits risk for the farmer, it reduces his or her status almost to the level of a hired laborer (Abell, 1968: 210).

Contract farming is of obvious benefit to the contracting firm. It helps to control the quality of production, reduces unit cost, and assures continuity of supply of products to be processed and retailed. It has also been one of the most dynamic factors in changing farming methods. The contracting firm not only supplies the most modern inputs and provides managerial supervision to the farmer, but also coerces the farmer into following the most profitable and productive techniques in producing commodities at a high standard through the use of strict sanctions provided for in the contract (Fortin, 1968: 365-5). However, only a few farmers, with proper previous training and skill are able to adapt successfully to contract farming and at the same time manage their farms themselves. The majority of integrated farmers, who lack the requisite ability, training, and skill, "progressively become nothing more than manual workers" (Fortin, 1968: 365) soliciting a contract almost similar to that of piece work laborers, by demanding to work at a set price for fixed quantities instead of assuming the risk of managing their farms. Contract farming thus creates an economic disparity in rural communities by favoring a few who can adapt to its demands and by proletarizing or eliminating other farmers who fail to do so. In other words, although vertical integration in the form of contract farming has been beneficial to agriculture in the sense that it has contributed to the rise of agricultural productivity by effecting a rapid change in farming methods, "it appears just as dangerous a solution for the individual farmer and for the rural class in general" (Fortin, 1968: 365).

Knowledge and Education

An important factor responsible for modernization of agriculture in Canada is the increased level of educational attainment by farm workers. Although the farm segment of the Canadian population continues to manifest the lowest level of educational attainment in comparison with the nonfarm and urban, the proportion of farm population with secondary or higher level of education has increased significantly over the last several decades. By 1991, over three-fourths of the farming population had at least attained a secondary level of education. There has also been a significant increase in the proportion of farmers and farm workers with nine or more years of formal schooling: from only less than one-fifth in 1941 to almost one-half in 1981. Agricultural training programs offered by the Manpower Department of the federal government in farm management, accountancy, and selection, use, and operation of farm machinery have helped farmers to develop skills needed to participate in modern farming activities.

The extension agencies also play an important role in helping farmers to continuously update their knowledge and skills in an environment of rapid technological change. Provincial governments are primarily responsible for agricultural extension in Canada, and the federal government for agricultural research. The role of agricultural extension service is to disseminate to farmers the results of experiment stations and university research which can be used to improve the farm and its productivity. It also serves indirectly as a sales promotion tool for machine, feed, seed, fertilizer, and chemical industries as it informs farmers of new developments in agrifood industries. Agricultural extension, however, is a two-way process. Feedback from the farmer is often disseminated through extension agents to the researchers in experiment stations, universities and industries.

The role of the federal government is relatively modest in agricultural extension. Agriculture Canada operates a limited extension program at each experimental farm and research station. Its Small Farms Development Program and Farm Credit Corporation have extension implications. The agrifood industries, such as feed, fertilizer, and pesticide manufacturers, combine their extension services. Machinery manufacturers are also making determined extension efforts.

It is difficult to determine the adequacy of extension and its contribution to modernization of agriculture in quantitative terms. It is estimated that about one-third of all farmers in Canada are reached directly by the extension service (Task Force on Canadian Agriculture, 1977, v.1, p. A: 219). While there are many farmers who utilize the extension information to their advantage, there are others who use it very little. Identification of factors responsible for variable use of information on and adoption of technological innovations in agriculture has been a major topic of rural sociological research in North America during the last three decades.

Research and Development of New Farm Practices

The modernization of agriculture depends upon the willingness of farmers to adopt agricultural innovations resulting from research. These innovations contribute to the development of improved machinery and equipment, improved varieties and strains of crops and livestock, techniques to control or eradicate plant and animal diseases, insects, and pests, and improved managerial and marketing techniques. The research and experimentation for these innovations take place in federal research centers and experimental farms, universities, and industries in Canada.

Agriculture Canada conducts more than one-half of the agricultural research in the nation and has established 47 research units of varying size and specialization in different parts of the country (Task Force on Canadian Agriculture, 1977, v.1, p.A: 218). The research conducted by these establishments includes: economic and social factors affecting agriculture, animal diseases, grain, seed, and feed quality, and cereal breeding. About 14 percent of the research activity can be attributed to provincial governments. Almost 30 percent of the agricultural research is conducted in the universities across the country. The research conducted by the agrifood sector has been rather modest (six percent) and is mainly concentrated on agricultural engineering, plant breeding, soil fertility, weed control, and animal breeding and nutrition.

It is an accepted fact that agricultural research is a vital component of the growth of agricultural productivity: "Indeed, without agricultural research, new and improved production technology, and farm management practices contributing to increased agricultural productivity might not have evolved" (Tung and Strain, 1987: 37). It is, however, very difficult to empirically measure the contribution of research to technological development and hence to the growth of agricultural productivity. It is not a simple task to isolate the portion of productivity growth attributable to technology resulting from research activities from the portion attributable to other factors, such as increased use of inputs and structural changes in farming. According to Tung and Strain (1987: 38), one way in which research makes a contribution to the increase in agricultural productivity is "by increasing output with a given set of inputs." Increase in the yield of wheat, for example, has been possible because of the combination of high-yielding varieties of wheat developed through research and increased use of fertilizer and chemicals. Another way in which research contributes to the growth of agricultural productivity is through changes in input mix, allowing for more efficient use of all inputs, such as real estate, labor, machinery, seed, feed, fertilizer, chemicals, and energy including fuel and electricity. Tung and Strain (1987: 40-1) used an empirical equation developed by Brinkman (1984) to "estimate the contribution of agricultural research and technology to production and productivity growth." They found that during the period between 1961 and 1980, the annual rate of growth of agricultural productivity was 1.66 percent; the contribution of research and its supporting services accounted for almost 40 percent of the productivity growth while weather and all other unspecified factors accounted for the remaining 60 percent. They also found a close positive relationship between the amount of money spent on research and rate of growth in agricultural productivity.

Studies on Diffusion of Agricultural Innovations

The process by which improved farming practices diffuse among farmers has been a major topic of rural sociological research in North America and elsewhere (Jones, 1967; Rogers and Shoemaker, 1971; Dasgutpa, 1989). Although there has been few such studies in Canada, the generalizations of the studies conducted in the United States are largely applicable in the Canadian situation as well. The studies on diffusion of agricultural innovations involve analysis of factors which affect the rate by which an innovation spreads within a community or social system. There is always a time lag between introduction of an improved practice in a social system and its adoption by farmers. The purpose of diffusion research is to determine the method by which this time lag can be minimized.

The process of diffusion has been defined as "(1) acceptance, (2) over time, (3) of some specific item — an idea or practice, (4) by individuals, groups, or other adopting units, linked, (5) to specific channels of communication, (6) to a social structure, and (7) to a given system of values, or culture" (Katz *et al.*, 1963: 238). In other words, seven specific elements constitute the total process of diffusion.

Adoption

Adoption is defined as a process by which a farmer integrates an innovation in his or her system of farming and continues to use it. Adoption is not necessarily a permanent behavior for a farmer may decide to discontinue an innovation after using it for some time by replacing it with a more improved practice or due to other reasons. Rejection, on the other hand, involves failure to adopt an innovation.

A farmer often goes through a mental process from first hearing about a practice to finally adopting it. This is referred to as the *adoption process*. The adoption process can be broken down into a sequence of five stages: (1) awareness, (2) interest, (3) evaluation, (4) trial, and (5) adoption. At the *awareness* stage, the farmer comes to know about the existence of an innovation but has little knowledge or interest about it. The farmer develops interest about the innovation and seeks more information about it at the *interest* stage. In the *evaluation* stage, the farmer obtains more information about the innovation, makes mental application of the idea, and weighs its merits and advantages in his or her own farming situation. The farmer uses the innovation on a small scale to learn about how to use the practice in terms of amounts, time, and conditions of application in the *trial* stage. *Adoption* is the stage of acceptance leading to its continued use. Although the stages involved in the adoption process are somewhat arbitrary in nature, they have been found to be useful and empirically valid by several studies (Beals *et al.*, 1957).

Innovation

An improved farming technique or implement, a new strain of seed, and a more effective variety of fertilizer are some of the examples of agricultural innovations. An innovation may not be of equal interest to all. Its appeal as a novelty arises in its particular function in reference to a particular social category or group. While it may be perceived to be useful by an individual or group in one context, it may be seen of no use at all by another individual or group in a different context.

Several characteristics of an innovation have been found to be significant in affecting its rate of diffusion: (1) utility or relative advantage, (2) compatibility, (3) complexity, (4) communicability, and (5) divisibility (Rogers, 1962). *Utility or relative advantage* is the extent to which a practice is viewed as an improvement over an existing idea or practice which it would replace. The perception of its relative advantage or lack of it is affected by not only economic but also by social, cultural or personal considerations. *Compatibility* is the relative consistency of an innovation with the existing values and patterns of behavior in a social system. Farmers may refrain from adopting a particular farm practice for religious reasons. Considerations of recognition and social status sometimes stand in the way of acceptance of an innovation. On the other hand, an idea which is compatible with the prevailing system of values and behavior spreads at a relatively rapid rate. *Complexity* refers to the ready understanding and application of an innovation by potential adopters. *Communicability* is the case with which the results of a practice can be observed and communicated. The use of farm chemicals for killing insects, for example, is not only simple to understand and apply but its results also can be observed and communicated with ease. But the application of a new technique in growing a crop spread over several stages may appear to be too involved for easy observability of its results and its rapid acceptance. *Divisibility* is the degree to which a practice can be tried on a limited basis before it is adopted. Some practices can not be accepted on a limited and experimental basis by farmers before making a final decision for adoption. Obviously there is a considerable perceived risk in adopting such a practice without testing its applicability on a limited basis. Consequently, practices which are divisible have a higher rate of diffusion than those which are not.

Although the acceptance of an innovation primarily depends upon its characteristics as *perceived* by potential adopters, there are at least three qualities which are *inherent* to it which might be considered as conducive to the diffusion of an improved practice or idea (Ryan, 1969: 145-6). They are: (1) relative simplicity or ease with which the practice is understood and used; (2) communicability or the degree to which the results of the innovation are observable to potential adopters; and (3) divisibility or the extent to which the practice can be used on a limited basis without involving any major commitment in its initial use.

Time Lag

There is always a time lag between the point when an innovation is introduced into a social system and its acceptance by most or all of its members. The diffusion rate of an idea or practice in a social system normally starts slowly, then it increases rapidly until about one-half of all potential adopters have adopted it after which the acceptance continues at a decreasing rate. If the total number of farmers adopting an innovation at successive points in time are plotted on a graph, an "S" shaped growth curve will be obtained. Although all diffusion curves are S-shaped, the specific shape of the curve varies according to the type of the innovation and the social system within which it spreads. If the innovation is characterized by the attributes which allow it to diffuse at a rapid rate, the curve will have a short tail, will show a steep rise, and will level off sharply. On the other hand, if the practice is characterized by qualities which stand in the way of its rapid diffusion, it will have a relatively long tail, will rise gradually, and will level off very slowly. Thus, the diffusion curves of two separate innovations may be distinctively different in the same social system because of their differing attributes. The diffusion curves of the same innovation may also be different in a modern and in a traditional social system. The curve will take a steep rise in a modern social system because of its rapid rate of diffusion but will rise slowly and gradually in a traditional social system because of the reluctance on the part of its members to accept new ideas (Bose, 1964).

Adopter Categories

Based on the time individual farmers take in adopting an innovation, they can be grouped into five categories on an "innovativeness scale" (Rogers and Shoemaker, 1971: 180-5). Innovativeness, in other words, is the measure of immediacy of acceptance of an innovation by a farmer after its introduction within his or her social system. Adopter categories are based on classification of farmers of a community on the basis of their innovativeness. These are: innovators, early adopters, early majority, late majority, and laggards.

Innovators are the first to adopt an innovation after its introduction, and normally constitute about 2.5 percent of all farmers in a social system. They are often characterized by the highest level of education, income, and social status, and their farms are largest in size and most highly specialized. They are cosmopolite in orientation in that they have wide social contact outside of their communities, and they often receive information on new farming practices directly from their sources of origin rather than from local extension agents. They are, however, considered as deviants by their neighbors because of their venturesomeness, and are rarely asked for advice and information.

The *early adopters*, who follow the innovators in adopting a practice and constitute about 13.5 percent of all farmers in a social system, are localistic in orientation. Although they have a higher education, income, and social participation, own larger and more specialized farms, and are more innovative, they are not too far ahead of the average farmers in any of the above characteristics. Because of their more intimate participation in the local affairs of the community, they are considered as opinion leaders, and are sought for advice and information by their neighbors.

The farmers categorized as the *early majority* represent 34 percent of all farmers in a social system, and are slightly ahead of average farmers in innovativeness, education, income, social status, farm-size, and specialization. They adopt new ideas deliberately and willingly but do not lead as is the case with innovators and early adopters. Another 34 percent of all farmers are characterized as the *late majority*.

They have about the same characteristics as the early majority but to a slightly lesser degree. They are skeptical in their attitudes toward innovations and adopt them after most of their neighbors, friends, and relatives in the community have done so.

The *laggards* represent the remaining 16 percent of the farmers in a community. They are often the oldest farmers operating small farms in a traditional manner, and are lowest in education, income, and social status and participation. Since they are distrustful of innovations, they are the last to adopt innovative ideas.

Channels of Communication

Information relating to new ideas and practices disseminate from their sources of origin to the ultimate users through various channels of communication. Communication channels not only inform but also influence farmers in their innovative behavior. They can be broadly divided into two types: formal and informal. The sources which are deliberately organized and developed to disseminate information can be described as formal. The formal sources may again be divided into impersonal and personal. The mass media, such as television, radio, newspapers, newsletters and bulletins, are examples of formal-impersonal sources. Extension agents working for government departments or agrifood industries represent formal-personal sources. Informal channels are invariably personal, and are exemplified by neighbors, relatives, and friends who often serve as important sources of information, advice, and influence.

The mass media are most effective for disseminating information quickly to a large population. It largely involves a one-way flow of information. The use of personal sources, both formal and informal, involves a two-way interaction between the giver and receiver of information, and these sources are most effective in persuading the potential adopter to develop a favorable attitude toward an innovation. While formal and impersonal sources are most effective in the awareness and interest stages of the adoption process, informal and personal sources are most frequently sought by farmers at the evaluation and trial stages (Table 3.4).

Farmers in different adopter categories also vary in their use of sources of information. Innovators, for example, often bypass extension agents in favor of originating sources and not infrequently learn about innovations before extension agents do. They have formal and informal contacts with scientists and other innovators from whom they seek advice and information (Lionberger, 1960: 54). Early adopters have the most frequent contact with local extension agents, and also read farm magazines and extension bulletins for information. Farm magazines, and friends and neighbors are primary sources of information for the early and late majority. Laggards almost entirely depend on friends and neighbors for agricultural information. In summary, impersonal and formal sources of information are more important for innovators and early adopters than for later adopters. Innovators and early adopters also use sources that are closer in contact with the source of origin of an innovation, and of a greater variety than later adopters.

TABLE 3.4

STAGES OF ADOPTION PROCESS BY RANK ORDER OF SOURCES OF INFORMATION

Stages of Adoption Process	Sources of Information
Awareness Information	1. Mass Media 2. Neighbors and Friends 3. Extension Agency 4. Dealers and Salesmen
Evaluation Trial	1. Neighbors and Friends 2. Extension Agency 3. Dealers and Salesmen 4. Mass Media

Source: Lionberger (1960: 32).

Information on agricultural innovations follow a "two-step process" in spreading from their sources of origin to the farmers in a community. Ideas flow from external sources such as the mass media and extension agents to the innovators and early adopters. The innovators and, especially, the early adopters disseminate the information to the average farmers. In disseminating information, the innovators and early adopters not only inform but also influence and convince the average farmers about the usefulness of an innovation, and thus play the most crucial role as opinion leaders in a community.

Social Structure

Although decisions to adopt agricultural innovations are often subject to individual action, they are never taken in a structural vacuum. All potential adopters belong to numerous social groups, and few decisions can be made without regard to other members of these groups. The groups to which farmers belong affect their innovative decisions directly or indirectly (Lionberger, 1960: 67-89).

Studies on influence of family on adoption of innovations, for example, have shown that such family values as aspiration of high educational attainment for children and high value placed on social status and social participation are positively associated with the adoption of innovations. A farmer's neighborhood or community of residence often has direct influence on his or her adoption practice because the established norms of these locality groups serve as guiding principles of individual behavior. The community differences in adoption are often not a function of differing personal characteristics of resident farmers but of differing social structures and normative standards of the communities (Dasgupta and Madan, 1978).

Cultural Factors

While the groups to which the members of a society belong constitute its social structure, their beliefs, values, knowledge, and customs comprise its culture. Although distinct in conceptual terms, culture and society are closely interrelated. What people do as members of social groups and how they do it are determined by culture; and culture, on the other hand, is a product of human behavior in social groups.

Cultural factors have been found to be of great importance in studies on planned or guided social change (Foster, 1962). The beliefs, values, and customs of a group of people often act as barriers or stimulants to sociocultural change. A new idea or practice may be rejected by a group of people if it is considered as a threat to their established patterns. But if the practice is seen to be useful to satisfy their perceived need, and is compatible with their existing customs and values, it will be accepted. Because of the interrelatedness of customs in a culture, social, political, or regional considerations often stand in the way of technological innovations (Arensberg and Neihoff, 1964: 68-84). Spanish-American farmers in Arizona, for example, rejected an improved high-yielding breed of hybrid corn because their wives found the corn too hard to grind and did not like the taste as well as the traditional variety. Religious conviction of Amish farmers stand in the way of their acceptance of modern agricultural equipment such as tractors and electrical pumps. On the other hand, thousands of farmers of North India accepted inoculation of their cattle against *hemorrhagic septicemia* after they found out that none of the few cattle which were inoculated died of the disease in the preceding year.

Several studies reported significant association between attitudes and value orientation of farmers and their innovative behavior. Those farmers who put emphasis on economic security, conservatism, and traditionalism almost invariably had a low level of adoption; those who valued material achievement, material comfort and science, and identified themselves with efficiency and self-reliance models manifested a high degree of innovativeness in farming activity (Lionberger, 1960: 90-95).

Diffusion Studies in Canada

Only a few studies have been conducted on the diffusion of agricultural innovations in Canada (for example, Verner and Gubbles, 1967; Verner and Mildred, 1966; Millard, 1965; Gubbles 1966; Wall, Vaughan, and Driver, 1985; Culver and Seecharan, 1986). The findings of these studies, however, are very similar to those conducted in the United States. A study on the adoption of innovations by dairy farm operators in the Lower Fraser Valley of British Columbia, for example, reported that "not all agricultural innovations introduced to farmers are accepted, some are accepted but later discontinued, and frequently there is a delay on the part of the farmer in reaching a decision to accept or reject an innovation" (Verner and Gubbles, 1967: 8). While some farmers passed through the stages of adoption process at a relatively quick rate, others took a long time. Two sets of factors were responsible for delay in adoption or rejection of an innovation. The first category of factors were related to the characteristics of the innovations themselves: relative advantage, compatibility, complexity, divisibility and communicability. The second category included such situational factors as insufficient capital, small size of operation, and inapplicability of the innovation to the farm.

A few socioeconomic characteristics, however, were found to be significantly associated with the level of adoption of farmers. Two personal characteristics, social participation and enjoyment of dairying, were positively related to adoption; and the number of years on farm and the number of children showed negative relationships. Of the economic characteristics, number of young dairy stock and family farm plus off-farm income were positively related, and income from other farming activities was negatively related, to adoption of innovations. Age and years of schooling were not related to adoption, but farmers who took agricultural courses in vocational education and enjoyed dairying showed a higher level of adoption than others.

Different sources of information were used at different stages of the adoption process. Mass media and commercial sources were most important at the awareness stage. Government and individual instructional sources were most used at the interest stage. At the evaluation and trial stages, personal sources of information were of paramount importance. Almost half of the rejections occurred at the awareness stage which meant that sources of information failed to motivate and arouse interest in the innovations. The decision to reject involved less time than the decision to adopt an innovation. The rate of discontinuance was higher among those farmers who passed through the stages of the adoption process within a short period of time which suggested that those who decided to adopt in a hurry were prone to abandon the innovation later.

SUMMARY

The three sets of factors which appear to have contributed to the modernization of agriculture in Canada are: decline in labor input; increase in technological and mechanical input; and a combination of several more or less quantifiable factors, such as increase in farm size and scale of operation, commercialization and specialization, changes in marketing structure, increase in farmers' education and knowledge, and development and diffusion of agricultural innovations. It is, however, difficult to discuss their relationships with agricultural development in simple "cause" and "effect" terms, for while they have made important contributions to the modernization of agriculture, they have been influenced by the forces of modernization themselves. Moreover, many of these factors mutually influenced each other which, in turn, contributed to the changes in agriculture.

CHAPTER FOUR

Changing Agriculture: Consequences

While the factors discussed in the preceding chapter are themselves interrelated in terms of cause and effect, a number of economic and social factors can generally be identified as consequences of changing agriculture in Canada. The rise of agrifood industry involving interdependence between agriculture and industry in production, distribution, and marketing of agricultural goods is one of the consequences. Secondly, because of the phenomenal increase in the average size of farms and capital requirements for farming, agriculture is no longer a principal occupation for the majority of rural people. Thirdly, the cost-price squeeze in agriculture and expanding role of agrifood industry in farming have contributed to the rise of farmers' organizations and agricultural cooperatives to protect the interests of farmers. Finally, the rural depopulation caused primarily by highly commercialized and mechanized agriculture has significantly affected the social structure and institutions in rural Canada.

AGRIFOOD INDUSTRY

One prevalent point of view regarding the role of agriculture in Canadian economy is that its functional importance is rapidly declining. Whyte (1968: 43), for example, notes that between 1938 and 1962 the contribution of agriculture to the national economy fell from 22 to only nine percent in terms of net value of production. He also points out that the declining role of agriculture in the national economy is further evident in the fact that a very small proportion of people work in agriculture, and are directly affected by agricultural activities. This is true indeed if one accepts Whyte's "narrow" definition of agriculture. Appleton (1973: 4-5) argues that the functions and activities carried on by the farm and agricultural community have radically changed with the modernization of agriculture. When the majority of the Canadian population were directly involved in farming and received a major share of the consumers' dollar, it was appropriate to evaluate the role of agriculture in the national economy in terms of number of farms, number of people working on farms, farm cash receipts, assets, liabilities, and net worth. However, in view of the fact that only a small proportion of the Canadian population today are involved directly in the production of agricultural commodities, and receive much less than one-half of the consumers' dollar, the conventional method of calculating agriculture's contribution to the national economy is inappropriate.

As has been noted in the preceding chapter, with the modernization of agriculture there has been a structural differentiation in farming activities. The modern farmer has become dependent on off-farm industries for the supply of seeds, feed, farm chemicals and equipment as well as for the storing and marketing of his or her products. According to Davis (1956: 187), "modern agriculture is inseparable from the business firms which manufacture production supplies and market farm products." The term "agribusiness" refers to this close interdependence between agriculture and industry which includes "the sum total of all operations involved in the manufacture and distribution of farm supplies; production operation on the farm, and the storage, processing, distribution of farm commodities and items made from them" (Davis and Goldberg, 1957: 29). Appleton (1973) uses the term "agriculture and food system" to refer essentially to the same phenomenon referred to by Davis as "agribusiness" in the United States.

The "core" of the *agriculture and food system* in Canada, according to Appleton (1973), is comprised of industries which have close economic relationships with agriculture in terms of direct sales and purchases: agriculture and farm input; food and beverage processing; tobacco; and pharmaceutical. These industries are responsible for about 28 percent of the gross economic activity in Canada. The core, in turn, has several other seemingly far removed industries tied with it through direct sales and purchases of at least 25 percent: fabricated and structural metal; paper container; glass and glass product; petroleum product; forestry; hunting and fishing; transportation; and clothing. These industries together with those comprising the core constitute the agriculture and food system which account for about 36 percent of the gross economic activity in Canada. If the contribution of the industries indirectly related to the agriculture and food system are also taken into consideration, the total economic contribution of the agricultural and food system comes to 42 percent. This emphasizes the massive importance of the agriculture and food system in the Canadian economy.

The term "agrifood industry" is often used to refer to the "core" of Appleton's agriculture and food system. The agrifood industry, for example, "includes primary agriculture (farming) and the many industries and services that support agriculture or result from it — animal feed producers, food processors, and food wholesalers and retailers ..." (Basillais, 1999: 46). It accounted for almost nine percent of the Gross Domestic Product of Canada in 1977, created close to two million or 13 percent of all jobs, and produced about $100 billion worth of goods and services. Primary agriculture was responsible for 30 percent of the total product while the remainder was produced by the allied food industries such as food and beverage processing and retail and wholesale industries of the agrifood service sector.

Primary agriculture bought $14.5 billion worth of inputs from seed, feed, and fertilizer industries and produced $28.7 billion worth of goods for processing and retail industries. It also produced $10.6 billion worth of basic agricultural products for export. Food and beverage processing industries, in turn, produced $68.5 billion worth of goods for retail and food services, and for export. Finally, the retail and food services were responsible for $98 billion worth of goods. The agrifood industry also constitutes a vital component of Canada's international trade. For example, in 1997 its exports accounted for $21.5 billion or seven percent of total Canadian exports and its imports were worth $14.2 billion or five percent of all imports. The agrifood industry thus generated an agricultural trade surplus of $6.7 billion which was equivalent to 26 percent of the total Canadian trade surplus in 1997.

The rise of agrifood industry epitomizes the modernization of agriculture in Canada. Agriculture has become a part of a highly industrialized system which is extremely efficient and productive in economic terms. Agricultural productivity has consistently increased in the decades since World War II, even though the numbers of farms and farmers, and the acreage under agriculture have declined. On the other hand, the rise of the agrifood industry has contributed to the widening economic disparity in Canadian agriculture, fierce competition within farming population for economic survival, and the proletarization or elimination of medium and small farmers as agrifood industries entered into contract farming with agriculturists, and coerced them into accepting most modern and profitable techniques. According to some writers (Janssen, 1982; Hightower, 1978; Abell, 1968: 210; Fortin, 1968: 364-5), the rise of the agrifood industry has not proved to be beneficial to the majority of individual farmers and the rural class in general. It has transformed the farms into "factories in the field" as "the countryside has fallen to few strongest hands, and the rest have been forced to flee to the cities or to remain behind in bitter poverty" (Hightower, 1978: 281).

FARMING AS AN OCCUPATION

Rural people in Canada are not necessarily farmers anymore. Not only a mere three percent of the Canadian population lived on farms in 1996, less than five percent followed farming as a primary occupation. Agriculture is no longer a strictly on the farm occupation. There is an increasing interdependence between agriculture and

FIGURE 1
AGRIFOOD INDUSTRY OF CANADA, 1997

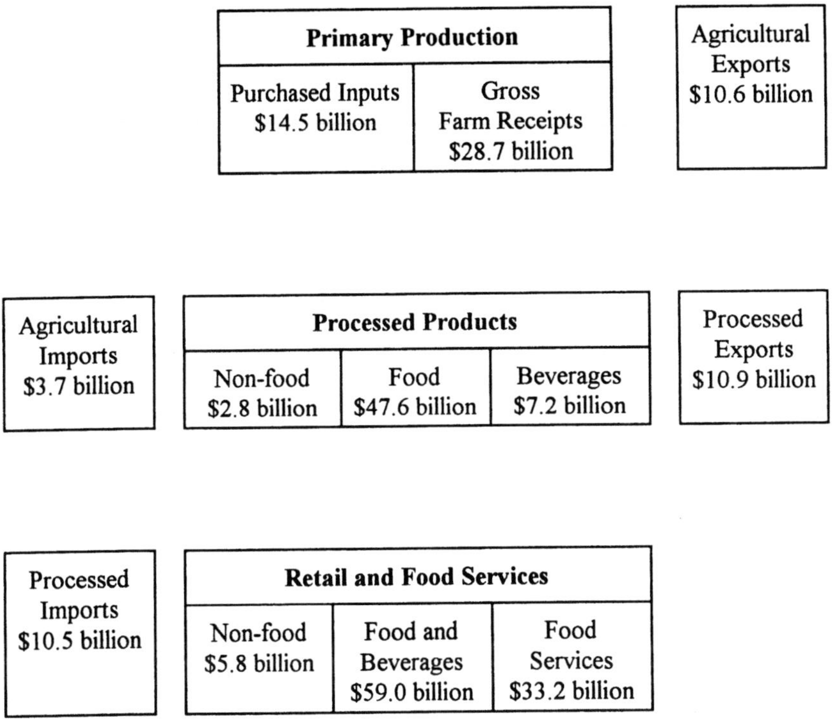

1.9 Million Jobs
13 Percent of Total Jobs
9 Percent of Total Gross Domestic Product

Source: Basillais (1999: 47).

industry as is evident by the rise of the agrifood industry. Farming today involves a great amount of capital, skill, and management ability to compete, and to be successful as a farmer. It is thus becoming increasingly difficult to take up farming as an occupation. The traditional process by which a farm boy starting as an unpaid family worker moved up to being a farm owner and operator has changed significantly. The rural and farm youth have become nonfarm job oriented in their occupational aspirations. The traditional pattern in tenure status with major concentration in the farm owners category is being replaced by a new pattern with concentration in the part owners category. Although the average farm income has shown a large increase over the last several decades, a small segment of successful farmers have benefitted most from agricultural modernization.

Agricultural Ladder

The term "agricultural ladder" refers to the stages a farm boy passes through before becoming a farm owner and operator. A young farmer traditionally began working as an unpaid laborer on the parental farm. In his late teens or early twenties he hired himself out to neighboring farms for a cash wage. When he saved sufficient funds to purchase a team and equipment, he advanced to the third rung of the ladder, and became a tenant or independent renter. The top rung of the ladder was attained when his earnings allowed him to accumulate enough money to make a down payment for a farm of his own. The majority of young farmers who started from the bottom of the ladder eventually became farm owners.

The agricultural ladder has undergone significant changes in recent years. Since a large proportion of the farm youth are now entering schools and colleges, the young unpaid family worker is no longer readily available. A young man is more likely to enter into a partnership with his father on the family farm rather than hire out as a wage hand. The eventual ownership of the farm may come to him not through the purchase of a separate unit but through gradual acquisition of control and ultimate inheritance of the family's farm.

Several factors have contributed to the changes in the agricultural ladder in recent times. Because of longer life expectancy and the tradition of not retiring from active farming until they are quite old in age, most farmers are not prepared to hand over farms to their sons when they are ready to take over. But farming requires such a substantial capital investment at present that it is almost impossible for young farmers to buy and start separate units. Productive farming units are so large and mechanized that the farm youth do not have the required capital to invest to buy farms and machines of their own. There is also a general decline in the availability of good land for farming purposes, and the price of agricultural land has risen well beyond the reach of young farmers.

A study of 352 Ontario farm families (Abell, 1961) has shown that the traditional agricultural ladder was still in practice in the Sixties but with considerable variations in the intermediate steps. It took about 17 years for a young man to climb the ladder, and most of the time was spent working for or with his father on the father's farm. Only four out of every ten farm-reared youth stayed on the farm and took up farming as a full or part-time occupation. One-fourth of these young men were on the first rung of the agricultural ladder. They were unmarried, lived with their parents, and worked on their parental farms for wages in cash.

About half of the young farmers were in the intermediate steps of the agricultural ladder with four different modifications: (1) about 14 percent were in their late twenties, unmarried, mainly worked on their parental farms, and also hired themselves out for a variety of unskilled nonfarm work; (2) four percent were in their early thirties, married, and worked for wages on a nonparental farm, and there was little likelihood that they would move up the ladder; (3) another four percent were in their late thirties, married, and were doing some farming as well as some type of unskilled nonfarm work; (4) 26 percent of the young farmers were about thirty years of age, married, and worked on their parental farms as full-time operators but had not yet achieved the full ownership status from their fathers.

The remaining one-fourth of the farmers' sons had reached the top of the agricultural ladder. The average age of these farmers was 35 years, and they completed an average of nine years of schooling. They were married, and lived on their own farms as full-time self-employed farm operators.

The agricultural ladder as a means of attaining the status of a farm owner is thus becoming nonoperative because of the highly commercialized and competitive nature of Canadian agriculture. It is virtually impossible at present for a young farmer to rise unaided from the position of an agricultural laborer to the status of a successful commercial farmer. The difficulty of obtaining the substantial capital investment required to start an independent unit has made inheritance the most important mechanism in attaining the status of a farm owner. Consequently, the rural youth are increasingly becoming nonfarm in their job orientation. Very few, especially those who did not grow up on a farm, take up farming as an occupation. A study of occupational expectation of high school seniors of Prince Edward Island (Dasgupta and Nagarajan, 1977: 11) has shown that only 12 out of 960 expected to be farmers. The majority expected to be professionals in business and in service industries.

Tenure Classes

The effects of modernization of agriculture are most evident in the changing pattern of tenure classes in Canada. Farmers can be categorized into different tenure classes according to their rights and relationships to agricultural land. There are several types of tenure classes in Canada which are of significance from a sociological point of view.

Farm Owner

A "farm owner" owns the farm he or she operates, normally has complete control over the farm, and usually enjoys full freedom in making farming decisions. However, owners of smaller farms, especially those who enter into contract farming with agrifood industries, have this freedom in theory rather than in practice. Indeed, under certain conditions of contract farming a farm owner may be reduced to the status of a laborer shorn of virtually all freedom in making farming decisions even though he or she may legally continue to own the farm. There is also a great economic disparity among farm owners in terms of size of the farm they own and their income. Owners of large farms have highly commercialized operations and high incomes. Owners of small farms often have subsistence-orientation in their farming activities, and have much lower incomes.

The proportion of farm owners has been falling steadily with increasing commercialization of agriculture (Table 4.1). Although the data for 1991 was not readily available, those from 1921 to 1981 show a consistent decline in the proportion of total land operated by farm owners (Table 4.2). Increases in the average size of farm and needed capital to expand operation to remain competitive in farming have contributed to this decline in the number of farm owners.

TABLE 4.1

FARM OPERATORS BY TENURE STATUS IN CANADA, 1941-1991

Year	Farm Owner		Part Owner		Manager		Tenant	
	Number	Percent	Number	Percent	Number	Percent	Number	Percent
1941	548,821	74.9	84,896	11.6	4,828	0.6	94,287	12.9
1951	481,447	77.3	89,226	14.3	7,435	1.2	44,983	7.2
1961	350,428	72.9	100,282	20.8	2,396	0.5	27,696	5.8
1971	251,066	68.6	95,862	26.2	--*	--	19,200	5.2
1981	201,657	63.3	96,857	30.4	--	--	19,847	6.3
1991	168,655	60.2	96,180	34.3	--	--	15,210	5.4

* Since 1971 census, a hired manager is asked to report whether the land operated by him or her is owned or rented by his employer and the farm is classified accordingly.
Source: Census of Canada.

Farm owners constitute a motley grouping of operators differing widely in size of farm owned, income, and sales of agricultural products. In his study of tenure patterns and commercialization of agriculture in Canada, Andarawewa (1969: 114) noted that almost 42 percent of farm owners grew less than $2,500 worth of agricultural products in 1966; and only 58 percent of all farm owners had commercialized operations. A large proportion of farm owners thus operated small and subsistent farms.

TABLE 4.2

PERCENT OF LAND IN FARMS BY TENURE OF OPERATORS
IN CANADA, 1921-1981

Year	Farm Owner	Part Owner	Manager	Tenant
1921	76.4	12.2	2.2	9.2
1931	64.4	22.0	1.3	12.3
1941	54.7	27.7	1.9	15.7
1951	54.0	33.9	3.1	9.0
1961	47.5	42.0	4.4	6.1
1971	44.7	49.7	--*	5.6
1981	39.2	54.0	--*	6.8

* A farm operated by a hired manager was classified under farm owner, part owner, or tenant category based on the information provided by the manager on the tenure status of his or her employer.
Source: Census of Canada.

Part Owner

A "part owner" owns some land and rents additional land to expand the operating unit. The proportion of part owners has been increasing consistently in Canada (Table 4.1). For example, from a little over one-tenth in 1921, part owners increased to over one-third of all farm operators in 1991 (Table 4.1). They cultivated over one-half of the total farm land in 1981 (Table 4.2). Part owners thus have greater net worth and larger operations than farm owners. Indeed, in his 1966 comparison of farm and part owners, Andarawewa (1969: 116) found that part owners operated a higher proportion of commercial farms than farm owners. The majority of part owners operate large and commercialized holdings, and are often the most successful farmers. Instead of locking in the highly needed, and difficult to obtain, capital in a fixed input by buying land to expand farming operation to remain competitive, they rent it from those who cannot afford to stay in the farming occupation any more. The trend of renting additional land to enlarge farming operation is further evident in the fact that from 1976 to 1996 the proportion of the total farm land rented by farm operators increased from 30 to 36.7 percent in Canada. Since 1901, this percentage has quadrupled.

Farm Manager

"Farm managers" operate farms for others for a wage or salary. The owners of the farms for whom they work may or may not live on the farm. Farm managers do not invest any capital of their own, but plan, supervise, and conduct business relating to farm operations on behalf of their owners. They may be entitled to a part of the net or gross income of the farm in addition to the wage or salary they receive.

Although the proportion of farms operated by managers doubled between 1941 and 1951, it declined by more than one-half in 1961, the last time farm managers were classified as a distinct tenure status category by Census Canada (Table 4.1). However, the proportion of land in farm manager-operated farms increased slowly but consistently and doubled between 1921 and 1961 (Table 4.2). This indicates that although the farms of this type decreased in number, they increased in size and operation. Most of the manager-operated farms today are highly commercialized and owned by nonfamily corporations.

Tenant

All the land operated by a "tenant" is rented. Ordinarily a tenant owns very little farming equipment, and has very little capital to invest in farming. Tenancy allows a person to operate a farm with limited capital. The proportion of tenants and land under tenant-operated farms, however, has been declining consistently in Canada (Tables 4.1 and 4.2).

Farm Laborer

"Farm laborers" work for wages in agriculture. Since they do not have any ownership or tenancy rights to the land on which they work for wages, they are not included in tenure classifications. Farm laborers, however, constitute an important segment of the agricultural labor force.

There are two major types of farm laborers: paid laborers and unpaid family laborers. The paid farm laborers are further subdivided into three types: year-round laborers, seasonal laborers and migratory seasonal laborers. A year-round farm laborer is normally paid a monthly cash wage, and is sometimes provided with boarding as well. Seasonal laborers are hired in peak periods of agricultural operations such as during the period of harvesting. They are hired for cash wages by the hour, day, or per unit of work. Although most seasonal workers are local non-farm people, a proportion of them are often migratory workers who come from outside of the local area, and sometimes from as far away as the Caribbeans, Mexico, and Europe under Foreign Seasonal Agricultural Workers Program in Canada.

The unpaid family laborers are family members of a farm operator who help them to perform various farm operations as a matter of duty. Working as an unpaid family laborer was the first rung of the agricultural ladder before a farm youth hired himself out as a laborer. Joint participation on the family farm was an important source of integration and solidarity of a farm family.

There has been a decrease in the total number of farm laborers in recent decades. Besides farm mechanization several other factors are responsible for the decline of the number of farm laborers: a decrease in the rural population base, low farm wages, and the more attractive wage and working conditions in urban areas. In recent decades different regions of Canada have experienced acute shortages of hired farm laborers especially those who have the needed skills.

While the total agricultural labor force has declined in numbers, there has been a gradual shift toward a greater reliance on hired labor. The number of hired laborers grew between 1951 and 1981 while the number of unpaid family workers declined (Table 4.3). Although unskilled laborers traditionally dominated the farm labor market, they are at present largely seasonal workers. The need for skilled laborers such as machine operators, mechanics, and skilled herdsmen has increased especially for the larger commercialized enterprises.

TABLE 4.3

NUMBER OF FARM LABORERS IN CANADA, 1951-1981

Type of Farm Laborer	1951	1961	1971	1981
		(in thousands)		
Hired Labor	218	263	246	347
Full-time	89	101	95	138
Seasonal	129	162	161	209
Unpaid Family Workers	506	306	258	279
Full-time	199	118	107	82
Seasonal	307	188	151	97

Source: Statistics Canada, Catalogue 71-001; Task Force on Canadian Agriculture (1977, v. 1, p. A: 55).

The number of unpaid family workers has decreased significantly since 1951 although it showed some increase between 1971 and 1981. The major factors responsible for this decline are increase in the number of farm youth continuing formal education and their changed occupational preferences, growth in nonfarm employment opportunities, and decrease in the average size of farm families. An increasingly large proportion of unpaid family laborers are female who are primarily farm wives helping out their husbands during busy seasons, and taking over some of the administrative functions of the farm such as bookkeeping and related tasks (Task Force on Canadian Agriculture, v. 1, p. A: 62).

Unfortunately, comparable data on hired labor and unpaid family workers, as presented in Table 4.3, are not available for 1991. The 1991 census, however, reported that there were 219,555 "farm laborers" including hired labor and unpaid family workers (Harrison and Cloutier, 1995: 1, 7). Assuming that the number included only "full-time" and excluded "seasonal" laborers, the total for 1991 (219,555) was almost the same as it was for 1981 (220,000). The trend of increased reliance on hired labor is further evident from the data in Table 4.4. During the twenty-year period between 1976 and 1996, there was a 14 percent increase in the number of farms using hired labor and almost a 70 percent increase in the average number of weeks of use of hired labor by each farm.

TABLE 4.4

USE OF HIRED FARM LABOR IN CANADA, 1976-1996

Use of Hired Labor	1976	1981	1986	1991	1996
Number of Farms Using Hired Labor	98,797	115,199	141,841	123,469	120,172
Percent of Total Farms Using Hired Labor	29.2	36.2	48.4	44.1	43.5
Average Number of Weeks of use of Hired Labor per Farm	37	39	40	43	53

Source: Statistics Canada (1996: 191).

Little substantive research on hired farmworkers of any kind is available in Canada. Wall's study on the offshore farmworkers (1987: 453-8), however, is one of the rare exceptions. As noted above, a proportion of seasonal farm laborers are migratory or "offshore" farmworkers who come to work in Canada primarily from Mexico and several Caribbean countries. The offshore farmworkers have few if any rights in Canada and are highly susceptible to exploitation. Since these farmworkers come to Canada without their families, Wall's study focussed on the extent to which they participate in the community life of their host country. She found that the offshore workers were peripheral to the ongoing life of their host community and their participation in the community was restricted to shopping in local stores. Residents of the host community tolerated their presence as long as they stayed "on the fields and in their place." The proximity of their residence to a community center and availability of transportation to the center and other community facilities also affected their participation in the community: "Some had the use of farm trucks and were close to the local centres while others were restricted to walking and had miles to go before reaching a store, post office, or any other community facility." Their ethnic background and linguistic abilities also influenced the quality of their lives in the host community. Because of their ability to speak fluently in English, the Caribbean workers enjoyed a greater acceptance than Mexicans who mostly spoke in Spanish. The latter, however, were considered harder workers than Caribbeans by local residents and "safer" to be around their wives and daughters. The residents of the host community thus used racial stereotypes in their attitudes toward offshore farmworkers and showed both fear and ignorance in relating to them. The relationship with the employer, according to Wall, had the most determining influence on the community participation of the offshore worker. Wall emphasized that regardless of "their ethnicity, how isolated the farm is, or how familiar the workers may be with Canada, their participation in the rural community life first requires the consent of their employers." The degree of the worker's participation in life beyond the boundaries of the farm depended primarily on the whims of their employers who not only gave the permission whether or not to leave the farm but also decided whether or not they could entertain visitors in their bunkhouses. Wall concludes by stating that offshore farmworkers "are the wage labour component of a capitalist style agriculture yet they get treated as an extension of the farm family, where one's personal freedom is determined by the whim and fancy of operators." Bolaria (1992) points out that the poor living conditions provided to the migrant agricultural laborers in Canada contributes negatively to their physical and mental health. Since access to offshore workers assures almost an infinite supply of labor, employers show very little concern for their laborers' health and well-being.

Corporation Farm

A major consequence of the modernization of agriculture has been the rise in the number of corporation farms. A "corporation farm" is a legally constituted business owned by share holders rather than by an individual or a family. The profits of a corporation farm are distributed among its shareholders according to the value of their share. The majority of these farms in Canada, however, are "family corporation farms." The shares of a family corporation farm are held by the members of a family. Many large farms which started as family-sized farms have now become family corporation farms, and are run according to business principles. "Nonfamily corporation farms" are generally operated by large processing or wholesaling industries or by retail chain supermarkets.

TABLE 4.5

FARMS BY TYPE OF ORGANIZATION, 1971-1991

Type of Farm	1971		1981		1991	
	No.	Pct.	No.	Pct.	No.	Pct.
Individual or Family	336,175	91.8	275,779	86.6	177,695	63.5
Partnership	21,019	5.7	29,534	9.3	78,024	27.9
Corporation	7,992	2.2	11,989	3.8	23,265	8.3
Family	7,081	1.9	10,742	3.4	19,230	6.9
Nonfamily	911	0.2	1,247	0.4	4,035	1.4
Other	942	0.2	1,059	0.3	1,059	0.4
Total	366,128	100	318,361	100	280,043	100

Source: Census of Canada.

In 1991 there were 23,265 legally constituted corporation farms representing slightly over eight percent of all farms in Canada (Table 4.5). The vast majority of them were, however, family (6.9 percent) rather than nonfamily (1.4 percent) corporation farms. An additional 78,024 farms were "partnership" farms involving two or more operators with or without written agreement among partners. Both partnership and corporation farms have been increasing in absolute and relative terms. The proportion of partnership farms, for example, increased almost five times, from 5.7 percent to 27.9 percent from 1971 to 1991. The proportion of corporation farms quadrupled from 2.2 percent to 8.3 percent during the same period. While the proportion of family corporation farms doubled, that of nonfamily corporation farms increased three times. Nonfamily corporation farms operated over one-fifth of all farm lands in 1981. It was feared by the Task Force on Canadian Agriculture (1977, v. 3: 62) that "we may be only a few decades away from a time when nonfamily enterprises are much more common than at present. Such enterprises are even now gradually expanding in numbers and influence while the number of family farms are declining rapidly." On the other hand, there are others (Lyons and Kraut, 1992: 593) who maintain that over 98 percent of total farms in 1991 were family-operated farms regardless of whether they were individual or family holding, partnership, or legally incorporated, and although "there has been major evolution in the characteristics of these units," family-operated farms have dominated the statistics and continue to do so. They insist that nonfamily ownership and control of agriculture have not changed significantly in recent years and they do not see forces likely to change this basic statistical fact in the near future.

Decline of Family Farms

Although the overwhelming majority of farms in Canada are indeed family-operated, they are vastly different from what is traditionally understood to be "family farms." It is difficult to define a traditional family farm in terms of physical size except to describe it as a unit which could be managed and operated by one family with major reliance on its own financial and labor resources. The farm family not only owned fully the farm it operated but also all of its other capital assets such as the buildings, machinery, livestock, and crops. A family farm relied primarily on unpaid family labor except in peak periods of agricultural operation when it hired some labor as well. Most farm management decisions were made by the male head of the family who also allocated daily tasks to other members, instructed them in the technical aspects of their work, and provided supervision and leadership to the various farm activities. The wife in a traditional family farm performed the role of mother and homemaker and sometimes assisted her husband in field work and making management decisions. Since the family farm operator was his own boss, he enjoyed the freedom of decision making relating to the management of his farm, lived primarily on the income he earned from farming, and relied solely on his own rather than borrowed resources. The family farm was thus largely a self-contained unit of production and consumption (Task Force on Canadian Agriculture, 1977, v. 3: 53-77).

The mechanization and commercialization of agriculture has made it more and more difficult for the family farms to compete and survive. As has been noted in the preceding chapter, between 1941 and 1991 over 8,000 farms, most of which were family owned and operated, went out of existence every year. While the number of farming units declined by almost two-thirds between 1941 and 1991, average size of farms increased two and a half times. Although the vast majority of farms are still family-operated, they are very different from traditional family farms in size, capital assets, productivity, and mode of operation.

The average size of the family-operated farms in 1991 was well over 400 acres. While 60 percent of the operators of these farms were Farm Owners, 34 percent were Part Owners. The operators of the latter category are aggressive commercial farmers who expand their operation by renting additional land.

Contemporary family farm operators have to use borrowed funds because the magnitude of their capital investments are much greater than those of their traditional counterparts. Very few operators of family farms today own their entire capital asset outright. They use more hired labor, and rely less on unpaid family workers. Their freedom of decision-making is substantially affected by production quotas established by various marketing boards, and by contract farming.

The two features in which the modern family farm deviates greatly from their traditional counterpart are: dependence on farm incomes and self-sufficiency. Almost 40 percent of family farm operators did some off-farm work in 1991, and 14 percent reported the equivalent of a full-time off-farm job. The mechanization of farm work allowed many of these farm operators to devote time in off-farm occupations. Because of the highly specialized and commercialized nature of farming today, a family farm is no longer a self-sufficient unit of production and consumption. A modern farm family, like its nonfarm and urban counterparts, buys most of the goods and services it consumes, including food, from outside.

The contemporary family farm is in the process of organizational transformation as well. It is showing an increasing trend toward partnership and incorporation in its organization. While partnership and share holding are usually confined to the family members, in some cases partners and share holders come from outside the family. Increased capital requirements for a viable farm unit, problems associated with intergenerational transfer of family farms, and changing tax laws are contributing factors to this trend of family-operated partnership and corporation farms (Task Force on Canadian Agriculture, 1977, v. 3: 56-8).

The family farm is highly idealized in North America for upholding three valued traditions: the tradition of man-land relationships, the tradition of democracy, and the tradition of efficiency in making most of one's resources. The decline of the family-sized farm is viewed by many people, including some sociologists and agricultural economists, with great concern who believe that the institution should be maintained. Indeed, a poll in the United States indicated that 71 percent of the rural sociologists and 41 percent of the agricultural economists surveyed maintained that the small family-sized farm operated by an owner should be the goal of American agriculture even though their productivity and income may not be as high as large commercial farms (Slocum, 1962: 297). The supporters of the family farm believe that it is an "efficient and dependable supplier of food," it "provides a superior form of social organization," "environmental problems and dependence on nonrenewable resources are minimized with a family farm structure," and its demise "threatens the viability of rural communities." The critics, on the other hand, maintain that the insistence to preserve an "unchanged archaic form" of farm organization is inconsistent with dominant North American values. They also argue that the concern that the family farm is disappearing is unfounded; modern family farms have adapted successfully to the changed conditions by mechanizing and enlarging their operating units (Brewster, 1958: 1606-7). The supporters of family farms, however, point out that a farm cannot grow indefinitely and still be called a "family farm." A large, mechanically operated family farm is not very different from a large commercial farm which is operated like an industrial plant (Sinclair, 1958: 1611).

Types of Contemporary Family Farms

Not all family-operated farming units today are large, mechanized, commercial, and highly productive. They exhibit a great heterogeneity in size, method of operation, productivity, and income. Indeed, a significant proportion of these farms continue to be of small scale in size, operation, and productivity, and earn a low income by today's standard. Although a degree of economic disparity always existed in Canadian agriculture, the causes of its recent accentuation can be attributed to the modernization of agriculture. While a small group of aggressive farmers has successfully enlarged the scope of their operations, and raised their productivity in phenomenal terms, the great majority of average farmers is far below this level of economic activity. Only a small minority of large scale operators represent the most modernized segment of the farming population, and is responsible for growing most of the food and fibre. The present-day family-operated farms may be broadly classified into four types: corporation, partnership, commercial, and subsistence.

Corporation Family Farms. Corporation family farms are the largest and most progressive of the family-operated units. In 1991, they accounted for over 80 percent of all corporation farms but only 7.5 percent of all family-operated farms. These farms evolve from partnership family farm types as they reach a certain size in terms of assets when problems related to tax and intergenerational transfer make incorporation desirable. By incorporating the business, a family corporation can take advantage of the tax benefits relating to depreciation and investment like other corporations. Secondly, a corporation family farm can be kept intact even after the death of its original owner by dividing the owner's stock rather than the farm among heirs.

Typically, an incorporated family farm is much larger than an average family farm. In 1981 it was three times as large in average size as its unincorporated counterpart.

The highly commercialized nature of these operations are indicated by the fact that 58 percent of corporation family farms had a gross sale of $100,000 or more in 1981.

Partnership Family Farms. A partnership farm is run by two or more operators as partners of a business with or without a written agreement. Most of these farms are family partnerships among members of families. They account for over one-fourth of the total number of farms in Canada and are increasing in number and size since 1951 (Task Force on Canadian Agriculture, 1977, v. 1, p. A: 63). Partnership farms in 1981 occupied about 12 percent of the total farm area, operated larger than average farms, and about half of them had a gross sale of $50,000 or more.

Commercial and Subsistence Family Farms. About 64 percent of all farming units in 1991 were individual or family farms which were neither incorporated nor partnership operations (Table 4.5). They, however, constitute a highly heterogeneous category in size, scale of operation, gross sale, and productive capacities. Based on their gross annual sale these farms can be classified into two general types: commercial and subsistence (Table 4.6). A "commercial" family farm earns a significant part of its income, if not all, by selling its products in the market. The income earned is used for living and invested in farming. The "subsistence" farms do not produce enough to maintain an adequate living standard from its income. A subsistence farm operator either relies heavily on income from off-farm sources or lives in poverty.

The commercial family farms with large gross sales ($50,000 or over) made up over one-third of all nonincorporated family farms in 1991 (Table 4.6). These farms occupied about one-half of the total farm area cultivated by family farms, and were more than twice the size of an average family farm. They represented the most modern and commercialized of all family farms, and accounted for over two-thirds of the total agricultural production in Canada.

The commercial family farms with medium gross sales (between $49,999 and $25,000) made up almost one-fifth of all family farms. Although not as well off as those of the large family farms, the operators of these farms cultivated larger farms than the average family farmer and produced enough to rely primarily on their farm income.

TABLE 4.6

FAMILY FARMS BY VALUE OF AGRICULTURAL PRODUCTS SOLD IN CANADA, 1981-1991

Scale of Operation	Value of Products (sold in dollars)	1981 No.	1981 Pct.	1991 No.	1991 Pct.
Commercial:					
Large	100,000 and over	21,947	8.0	29,929	16.8
	99,999 - 50,000	45,929	16.6	33,520	18.9
Medium	49,999 - 25,000	53,298	19.3	31,085	17.5
Small	24,999 - 10,000	54,215	19.7	34,359	19.3
	9,999 - 5,000	31,122	11.3	18,912	10.6
Subsistence:					
	4,999 - 2,500	25,632	9.3	13,256	7.5
	Under 2,500	43,636	15.8	16,634	9.4
Total		275,779	100	177,695	100

Source: Census of Canada.

Almost one-third (30 percent) were small and "limited resource" family farms with gross sales between $24,999 and $5,000. Limited resource farmers are generally serious farmers but "operate small units, and have resources meagre in quantity and quality — land, livestock, management, or capital — to earn adequate incomes from agriculture. Some have combined farm with nonfarm work, but many have remained wholely dependent on agriculture" (Trant and Birkman, 1979: 21).

The remaining 17 percent, designated here as "subsistence" farmers, had less than $5,000 in gross sales. An analysis of the income of all farmers who depended heavily on farming, but had a gross sale of less than $5,000 in 1974 has shown that 60 percent had incomes below poverty level (Task Force on Canadian Agriculture, 1977, v. 3: 61). These subsistence farmers operated small farms with outdated technology, were reluctant or unable to adopt change, relied heavily on the family garden as a source of food, and were oriented toward self-sufficiency like traditional family farmers. The majority of the farmers with a gross sale below $5,000 are, however, hobby farmers who often suffer net losses from farming and rely principally on substantive off-farm incomes (Trant and Birkman, 1979: 26).

A comparison between 1981 and 1991 shows that the proportion of large commercial farms grew by over ten percent during the ten-year period while those of medium and small scale operations declined somewhat (Table 4.6). There was a significant decline in the proportion of subsistence farms. Family farms, other than partnership and incorporated units, thus had to continuously increase their scale of operation to survive.

Off-Farm Work and Part-Time Farming

Off-farm income is increasingly becoming a major element of the total income of farm households. Conversely, farm income is contributing a declining proportion to the total income of farm operators. While, for example, in 1971 almost one-third of the total income of farm operators was derived from farming, in 1991 the proportion declined to less than one-fifth (Harrison and Cloutier, 1995: 49-52). Off-farm employment income accounted for over one-half (55 percent) of the total income of farm operators. The remaining 25 percent of the farm operators' income came from other sources, such as investment, government, and retirement income.

Operators of larger farms (those with gross farm receipts of $50,000 or more) earned a larger proportion of their total income from farming than those of smaller farms (gross receipts of less than $50,000). Indeed, three-fourths (75 percent) of the income of smaller farm operators was employment income while a little over one-third (38 percent) of the total income of operators of large farms came from off-farm jobs. More than twice as many smaller farm operators (49 percent) worked off-farm as operators of larger farms (23 percent): "This, of course, makes sense since the small farm is either supplementing the operator's major income, or is the 'other job' that supplements farm income" (Harrison and Cloutier, 1995: 48).

More than one-third (37 percent) of all farm operators did some days of paid off-farm work in 1991. Over one-fifth of all farm operators worked full-time — 190 days or more — off the farm for pay. While only nine percent of operators of larger farms worked full-time off the farm, the corresponding proportion for operators of smaller farms was 31 percent.

Part-time farming is thus becoming a significant type of economic activity in rural Canada. "Part-time farmers," defined as those who report at least 127 days of off-farm work in a year, increased from 20 percent in 1971 to 24 percent in 1981. The highest proportion of part-time farmers were found in British Columbia where one-half of all farmers did some off-farm work, and over one-third reported at least 127 days of off-farm work in 1981. Both Ontario and the Atlantic provinces also had large percentages (30 and 29 respectively) of part-time farmers working at least 127 days in a year off the farm.

In his study of part-time farming in Nova Scotia, Gunn (1978) found that an average part-time farmer was younger and operated a smaller farm than a full-time farmer. The part-time farmers, however, were largely concentrated in the same capital-size class ($24,950 - $49,949) as the full-time farmers although they had a somewhat lesser value of farm sales. Two-thirds of the part-time farmers had farm backgrounds but most of them acquired their farms in recent years through outright purchase, inheritance, receiving it as a gift, or some combination of these processes. The majority of them wanted to continue to do part-time farming and some planned to make it a full-time occupation either after retiring or quitting the off-farm job.

The productivity of the part-time farmers was lower than their full-time counterparts. Indeed, an average part-time farmer in Nova Scotia suffered a net loss in his farming operation. The low prices of the commodities they produced, insufficient production from capital invested, and inadequate financial return were the major problems associated with part-time farming. The high cost of machinery and equipment for a relatively small acreage, and insufficient time invested in farm work because of off-farm employment, were the major reasons for low productivity and lack of profitability in part-time farming. The debt load of an average part-time farmer, however, was low, indicating that he or she took little risk, made limited use of available technological resources, and subsidized farm work with an average off-farm earning of $13,585. It should be noted that part-time farming is not universally characterized by low productive efficiency and lack of profitability. Many part-time farmers, especially those of the west coast in Canada, operate modernized, productive, and profitable enterprises.

Two factors appear to have made major contributions to the increasing trend toward part-time farming: the cost-price squeeze in agriculture and the movement of people from the central city to the rural-urban fringe. The increasing cost of inputs has forced farmers to expand their operations, and use the most modern technology to raise productivity and maintain profitable income. Those who failed to expand their operations and utilize the latest available technology, either moved out of farming altogether or took a full-time or part-time off-farm employment and retained the farm as a modest enterprise. The movement of some urbanites dissatisfied with the city life to outlying rural areas has also increased the number of part-time farmers. These urban migrants to rural-urban fringe areas use their farms as a place of residence or a relatively small scale operation or both while they continue to earn most of their income from off-farm employment.

Barlett's (1986) categorization of part-time farmers in the United States can be applied to the Canadian part-time farmers as well. Barlett identified three types of part-time farmers: transitional farmers, standard farmers, and investors. Part-time farmers labelled as "transitional" originally devoted all their time to farming until difficult economic circumstances forced them to take work off the farm. The transition to part-time farming allow them to save the family farm or increase family income. They either reduce the size of the farming operation to be able to spend time in off-farm job or they maintain the size of the operating unit relying on family or hired labor. Those part-time farmers who are referred to as "standard farmers" by Barlett first established themselves in a nonfarm full-time occupation before taking up farming as a part-time activity primarily because it requires a large capital to enter full-time farming at present. Few standard part-time farmers will be able to fulfill their aspiration to be full-time farmers due to the difficulty of accumulating the large capital and resources they will need to invest. Finally, "investors" represent the type of part-time farmers who always had full-time jobs off the farm. Many investors got into farming after they inherited their farms. Investors are also often hobby farmers who are urban professionals with a fairly high income and are in farming not for an income but to enjoy certain aspects of agrarian life. Another group of part-time farmers who could be included in the investors group are the "back-to-the-landers" who are well-educated and relatively affluent urbanites attracted to agrarian lifestyles.

In view of the fact that less than one-fifth of the total income of Canadian farm families comes from farming, over one-third of farm operators have an off-farm job, and an increasing incidence of part-time farming, Lyons and Kraut (1992) argue that there is a growing need to pay attention to and emphasize the positive role of part-time farming in Canadian agriculture. They reject the conventional characterizations of part-time farming that it is inefficient and is a measure of the sorry state of agriculture in Canada. Part-time farming, they argue, plays a positive role in the welfare of farmers and the health and stability of agriculture. Since part-time farming and off-farm income are linked, opportunities and sources for off-farm employment and income must be improved: "the value judgement that family farming and full-time farming are the only 'goods,' and that part-time farming and off-farm income are 'bads,' are simply outdated." Part-time farming and off-farm income are thus firmly established as part of the structure of Canadian agriculture, and they must be treated on an equal basis with full-time farming for the purpose of framing agricultural policy.

Women in Agriculture

Women's role in agricultural production has generally been underestimated in official statistics in Canada. Reimer's study (1986: 152) of a small parish in Quebec suggests that official statistics on participation of women in the agricultural labor force "must be doubled or even tripled" even when a narrow definition of farm labor is used. If the definition of farm labor is broadened to include "errands, gardening, and conserving food" it is found that women spend only about 32 minutes less per day than men in farm work, and the labor expended by women contributes directly to the operation of the farm. Another study (Shaver, 1990) showed that total labor is equally divided between women and men when a broader definition of farmwork is combined with off-farm and domestic work. Although they expanded their definition of farmwork in a slightly different manner, Smith (1987) and Bollman and Smith (1988) came up with similar findings. Recognition of the linkage between the household and the farm thus brings women's contribution to the economic well-being of farms to the forefront.

Over one-fourth (26 percent) of all farm operators in 1991 were women. Most of the women operators, however, were found in two-operator farms generally operating their farms with their husbands. While only six percent of women were operators in one-operator farms, 44 percent of them were operators in two-operator farms, and nine percent in three-operator farms. In the vast majority of cases (84 percent) women were farming with their husbands while only one-third of men were farming with one or more female operators. This can be explained by the fact that a high proportion of men were associated with one-operator farms which accounted for two-thirds of all farms (Harrison and Cloutier, 1995: 20-1).

In summary, the traditional family farm has virtually disappeared although the vast majority of the farms in Canada are still owned and operated by family units. An average family farm at present is more than twice as large as the traditional family farm, and is a highly commercialized business. There is, however, a great variability among contemporary family farms in size of operation, use of modern technology, and productive capability. While a little over one-third of the family farms are very large and modern operations, some of which are run as partnership or incorporated enterprises, and produce most of the food and fibre in Canadian agriculture, small family farms continue to constitute a large proportion of the primary agricultural producers. The operators of small farms have limited resources, are unable to use the modern technology, and earn inadequate incomes. Although a proportion of these small operators are part-time farmers who have an adequate income from off-farm employment, those who depend primarily on farming activity often have an income below the poverty level.

According to one estimate (Task Force on Canadian Agriculture, v. 4: 13), if the present trend continues, Canada will be left with only 250,000 farms within a decade or two, less than half of which will be operated by full-time farmers dependent primarily on agriculture. These operators will dominate farm production, and run their enterprises as partnerships and small corporations.

The role of farming as an occupation in rural Canada has thus been radically transformed with the commercialization of agriculture and the decline of family farms. The number of farmers has decreased in phenomenal proportion as farming has become capital intensive and fiercely competitive as an economic activity. The dramatic rise in the needed capital and skill has made the agricultural ladder as a process of moving up from the role of unpaid family worker to that of a farm owner obsolete, and an increasing number of farm youth are turning away from agriculture. Among those who continue to depend primarily upon farming, fewer own wholely the farm they operate, and an increasing number partly own and partly rent their operating units as more and more farmers rent additional land to expand their operations rather than locking up desperately needed capital by buying land. The farmers are a disparate occupational group with a few large, aggressively competitive and efficient farmers dominating Canadian agriculture while a large number of small operators have been earning declining farm incomes, and are unable to maintain an adequate living according to the prevailing Canadian standard. Off-farm occupation is increasingly becoming a major source of income of farming households. Part-time farming is also on the rise. Indeed, it has been suggested that off-farm income and part-time farming have become firmly established in the structure of Canadian agriculture. Finally, women are playing an increasingly expanded and important role in agriculture. One-fourth of all Canadian farm operators in 1991 were women although the vast majority of them worked with their husbands or male partners in two-operator and three-operator farms.

RURAL COOPERATIVES AND FARMERS' ORGANIZATIONS

Although the modernization of agriculture has caused a phenomenal increase in output and productivity, the majority of Canadian farmers have failed to adjust to the rapidly changing agricultural technology, and have either been totally eliminated from farming or have maintained a precarious existence as farmers. Lack of the needed resources and capital, inability to use the most modern technology, and the cost-price squeeze have contributed to declining returns from farming and increased reliance on off-farm sources of income for a large majority of farmers. Farmers, as a result, have attempted to protect and promote their economic and political interests collectively by joining cooperatives, commodity groups, and farmers' organizations in increasing numbers.

Rural Cooperatives

A "cooperative" is a voluntary association based on formal cooperation among its members who pool their resources and work collectively toward the goals which they cannot attain individually on their own. The growth of the cooperative movement in Canada has been largely due to farmers' struggle to improve their bargaining position. Consequently, the greatest development of cooperatives has been in rural and agricultural areas, particularly in marketing agricultural products and purchasing farm supplies.

Cooperatives are based on the principles of open membership; democratic control; limited return on capital; and distribution of surplus according to patronage. Membership to cooperatives is open to all regardless of race, creed, class, color, or political persuasion. They are democratically controlled in that each member has only one vote irrespective of the number of shares the member holds. Capital invested in a cooperative receives return at a modest rate as set in its annual meeting. The profits are distributed among members in direct proportion to the amount of patronage they give to the cooperative. A portion of the profits is always kept aside for a reserve fund to cover depreciation and unforeseen difficulties, and for educational purposes. A cooperative is an organization of users, not investors, whose primary goal is to serve the interests and needs of its members rather than to attract and serve customers, as is the case with a business organization.

Cooperative Movement in Canada

MacPherson (1982) identifies five stages in the evolution of the cooperative movement in Canada. The movement started in the period between 1850 and 1900 as a response to the rapid technological change caused by the first industrial revolution. Farmers sought to protect their interests by organizing cooperatives at local levels in the rural areas; but most of these early experiments which were largely spontaneous but lacked centralized structure and direction, and effective leadership, faded away.

The cooperative movement achieved its first full measure of stability between 1900 and 1920. These two decades were a period of extensive growth and industrialization. The prairies were opened up for settlement and cultivation, and mining, lumbering, and hydroelectricity rapidly expanded the northern frontier. The majority of farmers found it very difficult to keep pace with the scientific and technological advances made in agriculture. The cooperatives provided them with an organizational means to deal with the rapidly changing economic circumstances. Several marketing cooperatives were organized in different regions of Canada. The success of marketing cooperatives encouraged the growth of numerous consumer and financial cooperatives as well. The Cooperative Union of Canada which came into existence in 1909 provided a centralized structure and leadership to these cooperatives.

The cooperative movement went through the third stage of its evolution and experienced both successes and failures in the decade between 1920 and 1930. This was a period of rapid urbanization and further transformations in agriculture due to near universal use of electricity on the farms and in the rural homes. "Pooling," a new approach to cooperative marketing, made great advances not only among grain growers but also among the producers of other agricultural commodities. Pools were based on political neutrality and effective bargaining in the market place, and on localism rather than centralism in organizational emphasis. The contradictory nature of the cooperative movement in the 1920s is emphasized by the fact that while localism and regionalism rather than centralism were dominant forces, the movement also gained a measure of national strength for the first time during this decade.

In the fourth stage of its evolution, the Canadian cooperative movement took on a new life in the midst of the economic depression of the 1930s. The pools and marketing cooperatives suffered disastrous financial crisis because of the extremely fluctuating and unpredictable nature of the market place and commodity prices, and had to be bailed out by government assistance. Although the pools gradually recovered from the crisis, the experience convinced many farmers that changes in marketing techniques was essential and that the marketing board structure was a better means of marketing commodities. Provincial marketing boards came into existence, often with the assistance of cooperative organizations, during the later part of the decade.

The economic hardships of the 1930s gave rise to a variety of new cooperative endeavors and strengthened the existing efforts. Fishermen's cooperatives flourished; Credit Unions promoted by Catholic activists in the Maritimes spread westward; and varieties of consumer cooperatives came into existence in different parts of the country. In other words, "Canadian responses to the Depression was an extensive, all embracing attraction to cooperative solutions." The cooperatives in the 1930s, however, were more interested in financial gain rather than any ideological vision.

This trend became even more apparent from 1935 to 1955, the final stage of evolution of the cooperative movement. The Second World War radically altered the Canadian economy. Agriculture along with other primary industries became active again after a decade of stagnation which meant rapid and prosperous expansion of cooperatives in these sectors. An integrated financial system was developed for the cooperative movement during and after the war. The Canadian Cooperative Credit Society was organized in 1952 to provide support and leadership to the credit unions developed in nearly all provinces.

The social and economic changes following the war years began to affect the objectives of the cooperative movement. The ambitious vision of an integrated financial system for cooperatives proved too difficult to realize. The agrarian desire to reform society through cooperatives disappeared. The movement has since taken on a cautious and less controversial stance by defining its role as a third force in the economy; an alternative to capitalism, on the one hand, and public ownership, on the other. It has limited its activities to specific areas, particularly in agriculture, and aims to function as a cohesive whole. Although the movement does not arouse as much passion in the minds of its members as it used to, cooperatives have become a major force in agriculture. Their growth in number, membership, volume of business and assets have been impressive in recent decades.

Growth of Cooperatives

Cooperatives may be classified into six major types according to their functions: marketing, supply, production, service, fishing, and consumer.

Marketing Cooperatives. As their name implies, these cooperatives market primarily farm and agricultural products using the latest technology. Marketing cooperatives vary widely in size, volume of business, and commodities they specialize in marketing. Although their number remained more or less the same between 1990 and 1997, and their membership showed significant decline, their volume of business almost doubled (Table 4.7). Marketing cooperatives thus represent “a major economic force with a combined volume of business of more than 16.1 billion" and marketing "nearly $13.5 billion in agricultural products in Canada and abroad" in 1997 (McCagg, 1999: I). The market share of agricultural marketing cooperatives varies between 62 percent for dairy products and four percent for egg production.

TABLE 4.7

GROWTH OF COOPERATIVES IN CANADA, 1990-1997

Areas of Growth	Cooperative Type	1990	1995	1997
Number	Marketing	149	162	163
	Supply	220	247	257
	Production	380	461	522
	Service	3,069	3,904	4,038
	Fishing	58	56	50
	Consumer	604	582	582
	Total	4,480	5,412	5,612
Membership (in thousands)	Marketing	302	209	203
	Supply	431	347	360
	Production	22	31	34
	Service	810	932	655
	Fishing	10	8	7
	Consumer	1,845	2,977	3,439
	Total	3,422	4,504	4,698
Volume of Business (in million dollars)	Marketing	8,982	11,818	16,141
	Supply	2,075	4,002	3,098
	Production	459	778	876
	Service	1,024	1,701	1,400
	Fishing	185	165	134
	Consumer	3,193	6,382	7,128
	Total	15,918	24,846	28,776

Source: Sullivan (1993: 4); Gagne and McCagg (1997: 1); McCagg (1999: 1).

Supply Cooperatives. Supply cooperatives sell to their members and producers their needed agricultural supplies: fertilizers and chemicals, animal feed, farm machinery, seeds, and other inputs including petroleum products and building materials. Like marketing cooperatives, their number showed only a modest increase and the size of their membership declined between 1990 and 1997. Although the volume of their business doubled between 1990 and 1995, it went down by about 25 percent in 1997 (Table 4.7).

Production Cooperatives. Production cooperatives include "feeder, grazing, agricultural machinery, artificial insemination, forestry, and handicraft cooperatives" (McCagg, 1999: II). The number of production cooperatives as well as their membership have shown consistent growth between 1990 and 1997 and the volume of their business grew over 52 percent. Cooperatives in Quebec and Alberta accounted for more than 80 percent of the total volume of business of these cooperatives with Quebec deriving its revenue from the sale of forest products and Alberta from the sale of cattle and hogs (McCagg, 1999: II). About one-half of the production cooperatives are grazing cooperatives located in the Prairie provinces as also most of fodder, feeder, farm, and machinery cooperatives. Artificial insemination cooperatives are evenly distributed between Western and Atlantic provinces whereas the wood-cutting cooperatives are located mainly in Quebec (Sullivan, 1982).

Service Cooperatives. Service cooperatives make up the largest number of cooperatives in Canada accounting for over 70 percent of all cooperatives. While they consistently grew in number from 1990 to 1997, their growth in membership and volume of business tended to fluctuate by showing increase in 1995 over 1990 figures and decline in 1997. Service cooperatives include health insurance, housing, child care, health care, community development, funeral, water supplying, rural gas and electric, recreational, and transportation and communication cooperatives. Housing cooperatives were the largest in number (2,054) and membership (113,000) in 1997. Although there were only 51 funeral cooperatives, mostly in Quebec, they had a membership of 136,000. Most of the 210 natural gas and electric cooperatives were located in Alberta with over 119,000 members. There were 427 child care cooperatives in Canada with 29,400 members found in greatest numbers in Ontario, Saskatchewan, and Manitoba. Community development cooperatives "foster local-level leadership skills and full grassroots participation in innovation and solving socio-economic problems" (McCagg, 1999: III). The majority of 212 community development cooperatives reporting 12,027 members in 1997 were located in Saskatchewan. There were 169 water supplying cooperatives in Alberta, Quebec, and Manitoba; 288 recreational cooperatives, most popular in the West, particularly in Saskatchewan; and health care and transportation cooperatives found across Canada.

Fishing Cooperatives. Fishing cooperatives have been declining in number since the 1970s and the trend continues in the 1990s as well. From 58 in 1990, their number came down to only 50 in 1997 showing corresponding declines in membership and volume of business (Table 4.7). The fishing cooperatives in the Maritimes market the bulk of their fish to the United States and West Indies, and in recent years have started to exploit new markets in Japan, Europe, and the Middle East. The fishing cooperatives in British Columbia cater to the expanding export market in Japan.

Consumer Cooperatives. Consumer cooperatives rank immediately below service cooperatives in total number but rank at the top of all cooperatives in the total number of membership and volume of business. Although their number declined slightly in 1997 from that of 1990, their membership number almost doubled and the volume of business more than doubled during the same period. At the local level, consumer cooperatives vary greatly in size and sophistication — from small buying clubs to large supermarket retailers and wholesalers. Consumer cooperatives were found mostly in the Western (335) and Atlantic provinces (169) with sales of $2.8 and $1.1 billion respectively. There are consumer cooperatives in the rural regions of Ontario and Quebec as well although not as large and influential as in the Western and Atlantic provinces. There were 50 "direct charge" cooperatives in 1997. These cooperatives provide goods to their members at cost and charge them a weekly service fee to cover operating costs. There are over 100 student supply cooperatives in Quebec supplying from pens and pencils to course textbooks.

Credit Unions

Credit unions also form an integral part of the cooperative movement in Canada. In Quebec, credit unions are known as *caisses populaires*. Credit unions first came into existence in Quebec before it spread in English-speaking Canada and in the United States. Beginning with the creation of the first *caisse populaire* in Lévis, Quebec, in the year 1900 by Alphonse Desjardins, *caisses populaires* and credit unions have given rise to one of the most active national cooperative movements in the world. The first financial cooperative outside of Quebec in Canada was established in Ottawa in 1908, Civil Service Savings and Loan Society. They subsequently spread to the Prairie provinces to meet the financial needs of farmers during the Great Depression of the 1930s. During the next three decades they grew significantly, especially in Ontario and British Columbia, in response to the growing need for consumer financial services.

Credit unions have traditionally provided financial services to their members in regions, especially rural, where commercial lenders failed to provide these services. They are most active in Quebec, Saskatchewan, and British Columbia. Credit unions are owned and controlled by their members and are based on cooperative principles: each member has one vote regardless of the size of deposit or share capital held. As local cooperatives, credit unions offer financial services only to their members, and are organized at the level of a geographic, religious, ethnic, or occupational community. Originating as small, volunteer operated associations to serve medium and low income members, credit unions have grown into large and highly professional financial institutions since 1930. There were only 266 credit unions in 1930 with a total membership of 46 thousand and a total asset of $11 million (Cooperative College of Canada, 1982: 68-9). In 1998, there were 820 credit unions and 1,351 caisses populaires, with some 10 million members (33 percent of the Canadian population) and a total asset of about $110 billion. In 1997, 26 percent of Canadians used a credit union or caisse populaire as their primary financial institution. They are an important source of financing for small and medium-sized enterprises and the vast majority of their loans are under $1 million. Credit unions are strongest in Quebec and Saskatchewan followed by British Columbia. Membership is highest in Quebec, where almost two-thirds of the population belong to *caisses populaires*.

Organizational Structure

The organizational structure of cooperatives in Canada is of two basic types: federal and unitary. Credit Unions and retail cooperatives are organized in a federated structure. Local associations are federated at the provincial or regional levels, and regional units are federated into a single national organization such as the Canadian Cooperative Credit Society. Many of the marketing and consumer societies, such as the wheat pools, and dairy, trust, and insurance cooperatives are organized into large unitary structures which operate at local, provincial, and regional levels.

One of the unique features of the cooperative organizations in Canada is that several organizations at the national level with overlapping roles and membership draw together cooperatives of a wide range of types at the regional, provincial, and local levels into a single movement (Cooperative College of Canada, 1982: 83-4). While each of these organizations has a distinct role to play, they also maintain a relationship of close cooperation and coordination. The Cooperative Union of Canada, for example, looks after the interests of cooperatives by representing them at different levels of government policy and decision-making. The Canadian Cooperative Credit Society provides a national financial facility for credit unions and cooperatives. The Cooperative College of Canada performs a variety of roles in the educational area.

With growth and modernization, the cooperatives of Canada have become complex and bureaucratized. As in any large formal organization, the members delegate their authority to a board of directors at the annual meetings. The board of directors hires a manager who has the responsibility of carrying out and administering policies and decisions laid down by the board. Thus, the decision-making in modern cooperatives is left to experienced, knowledgeable, and skilled personnel rather than directly in the hands of members of the cooperatives. While such organizational transformation was necessitated by their steady growth and increasing complexity, and the need to compete effectively in the market place, the modern cooperatives have become structurally remote from their members "who lack an informed commitment to its meaning and purpose." A widely shared perception is that modern cooperatives behave more and more like their competition, and have developed an organizational style in which membership does not fundamentally matter (Cooperative College of Canada, 1982: 145). The cooperatives have become commercial enterprises with the primary goal of making profits for its members rather than acting as a medium to be used by a social class such as farmers to regain control of their destiny. Because modern cooperatives attempt to attract members through economic incentives alone by making their goal to be economic efficiency and profit, they have become bureaucratized and centralized, and have taken away real control from the hands of their members (Fortin, 1986: 356-60).

Cooperatives, nonetheless, have a crucial role to play in a changing rural society. Since they are both "market-oriented businesses and institutions designed to be responsive to the needs of rural people," they provide a link between decision-making of two kinds: "the 'calculative rationality' of the market, and the 'value-based rationality' of the community" (Fairbairn, 1997: 105). Cooperatives thus attempt to reconcile the conflict between "market pressures and community expectations" by internalizing them. As one of the central institutions of rural Canada, cooperatives must adapt to the changes and restructuring of the rural society. Indeed, for the last several decades they have been adapting to rural depopulation, increasing farm size, centralization and regionalization of rural services, and an increasing feeling of powerlessness among the rural people. Fairbairn (1997: 105) suggests that since cooperatives are controlled by rural people, "they also can act as agents of positive change, promoting the kinds of restructuring that are in the best interest of rural communities." Communities have the potential to create and employ new kinds of cooperatives as tools for adapting to and shaping economic change.

Commodity Organizations

Commodity organizations are formal groups which serve the interests of farmers growing commodities of specific types. They serve as trade associations in identifying problems, negotiating solutions, and providing an overall coordination to promote the interests of the growers of particular commodities. These associations have become quite numerous in recent years, and are active in lobbying and promoting legislation desired by their membership. Some of these commodity groups are: Canadian Cattlemen's Association, Dairy Farmers of Canada, Canadian Seed Growers Association, Canadian Fruit Wholesalers Association, and Canada Grains Council. Some commodity organizations operate on a national basis, work with different branches and departments of government, and carry on fieldwork to collect information relevant to promote the interests of their members (Task Force on Canadian Agriculture, 1977, v. 1, p. A: 265).

Farmers' Organizations

Most farmers in Canada belong to one of two major farmers' organizations: the Canadian Federation of Agriculture and the National Farmers' Union. The Canadian Chamber of Agriculture, later renamed the Canadian Federation of Agriculture, came into existence in 1935, and has acted as the voice of the Canadian farmers over the last five and a half decades. In recent years, however, its leadership among Canadian farmers has been challenged by a rival organization, National Farmers' Union, which was established in 1969. The history of farmers' movement in Canada, however, is at least one hundred years old, and several farmers' organizations came into existence since 1872 in response to the problems and crises experienced by the Canadian farmers (Wood, 1975).

Farmers' Movements in Canada

There was no organized farmers' movement in Canada until 1872. Local agricultural societies supported financially by provincial governments to organize agricultural fairs and exhibitions; and farmers' clubs, formed to discuss social, economic, and political issues, and create a bond of unity among farmers at local and community levels, were the only forms of organized farmer groups in existence until the Grange movement entered into Canada in 1872 from the United States.

The Grange movement entered Canada by way of the New England States, first into Quebec in 1872, and then spread quickly westward to Ontario, and to the Maritimes in the east. The movement, however, started to decline after 1897 because of its "questionable" business methods, secrecy of procedure, and alleged attempt to class dominance. The Grange all but vanished from the Canadian scene during the early decades of the present century.

Like the Grange, the Order of the Patrons of Industry also originated in the United States and spread across Canada by 1894. While the Grange was politically neutral, the Patrons were a political and economic association of farmers which also made provisions for the industrial workers to join in. The growth of the Patrons' movement in Canada was faster than the Grange but it withered away even at a faster rate by 1898.

In 1902 the Ontario Farmers' Association was organized as a politically neutral organization to fight for the cause of farmers in the Province. It, however, failed to attract a large membership, and ultimately merged with the Grange in 1907 after a compromise was reached that participation in Grange rituals would be optional.

The period between 1898 and 1912 is marked by the rise of the grain growers' movement in western Canada. Inadequate transportation facilities for marketing grains, monopoly of the ownership of grain elevators by a few large syndicates, and the alleged unfair practices of the elevator companies and the Canadian Pacific Railway in transporting grains were the major factors which contributed to the rise of the region-wide movement. The Territorial Grain Growers' Association of the Northwest Territories and the Manitoba Grain Growers' Association together started the Grain Growers' Company in 1906. It was owned and controlled by farmers. Similar farmers' organizations came into existence in Alberta and Saskatchewan as well. In 1907, the Interprovincial Council was established in western Canada. The Council proved to be so useful to the Western grain growers that a proposal was made in 1909 that the Council be expanded on a nationwide basis and the Dominion Grange be included under it. The Canadian Council of Agriculture was established in the same year the goals of which were to stimulate organization among farmers, collect and disseminate information relevant to farmers' interests, make proper presentation of farmers' interests at various political forums, and establish farmers' cooperatives.

The decline of the Grange created a large void in farmers' organization in Ontario until a new association, United Farmers of Ontario, was formed in 1914 with various existing farmers' groups in the province affiliated under it. The organization was expected to be equivalent in goals and activities to those already operating in the western provinces, and it aligned itself with the Canadian Council of Agriculture two years later. It achieved startling successes at the provincial polls in 1919 which resulted in the formation of a Farmer-Labor government in Ontario.

Farmers' movements in the Prairie provinces and Ontario inspired the organization of class conscious and politically motivated farmers' groups in the three Maritime provinces as well. United Farmers of New Brunswick was formed in 1918 and was affiliated with the Canadian Council of Agriculture two years later. The United Farmers of Nova Scotia was established and affiliated with the Canadian Council of Agriculture in 1920, as was the United Farmers of Prince Edward Island. In 1922, the three Maritime Associations were united for economic and financial reasons.

The Grange movement in Quebec was mainly limited among the English-speaking farmers although the Order of Patrons of Industry attracted many French-speaking farmers as well. In 1919, Les Fermiers-Unis de Quebec (The United Farmers of Quebec) was formed by the farmers of French origin. It was admitted into the Canadian Council of Agriculture in 1923. From the very outset the organization had political aims and put independent farmer candidates in several provincial and federal elections.

British Columbia remained outside of the influence of agrarian movements until 1917 when the United Farmers of British Columbia was formed. The organization, however, showed considerable hesitancy in committing itself to any independent political action, and preferred to confine the farmers' movement in the province to economic matters. It, however, maintained friendly relations with other farmers' organizations in the country through the Canadian Council of Agriculture.

The farmers' organizations in Ontario and the Prairie provinces took an active role both in provincial and federal politics in the early 1920s in promoting the interests of farmers and agriculture. The United Farmers of Ontario entered the provincial election in 1919, won the largest number of seats in the provincial legislature, and formed a Farmer-Labor government in coalition with the Labor party with one of their own leaders as the Premier although the government fell three years later due to internal discord and conflict. The United Farmers of Alberta contested the provincial election in 1922 in which three-fourths of its candidates were elected, and a farmer was chosen Premier of the government which was formed. Similar political success was achieved by the United Farmers of Manitoba in the 1922 provincial election when they gained a majority in the legislature. The farmers' organizations in the Maritimes were also successful in getting several of their candidates elected to the provincial legislatures in the early 1920s.

Electoral successes in provincial politics led to the formation of a political party in the federal arena under the auspices of the various farmers' organizations in 1920 called, the National Progressive Party. It formed a distinct caucus in the Parliament of Canada with elected independent agrarian members, and successfully contested three by-elections, and in 1921 it elected 65 of its candidates to the Parliament. The Progressive Party was thus able to attain strategic position in the Parliament by influencing the actions of the Liberal government. Internal dissensions and the formation of a "ginger group" within the party led to the sharp decline of its representation in the 1925, 1926, and 1930 general elections, and it failed to elect any member to the Parliament in the election of 1935 (Griezic, 1975: VI-XXII).

Both the Canadian Council of Agriculture and provincial organizations experienced a dramatic decline in their membership in the mid 1920s. Many provincial groups merged with the other political parties, went back to their former autonomous political affiliation, or simply went out of existence. The Canadian Council of Agriculture ceased its operation in 1932.

The severe postwar agricultural depression from 1920 to 1925, technological change in agriculture, and decline of the number of farms were major contributing factors to the setback of farmers' movements from 1920 to 1925. Farmers' interest in political action revived in the depression era when, in 1932, the Cooperative Commonwealth Federation (CCF) was organized as a loose federation of provincial affiliates. In Saskatchewan it became the official opposition in 1937, and seven years later formed the provincial government.

The economic hardship suffered during the depression years inspired the farmers to regroup in economic rather than political lines. The void created by the exit of the Canadian Council of Agriculture in 1932 was filled by the Canadian Chamber of Agriculture established in 1935. It was renamed the Canadian Federation of Agriculture in 1939. The goals of the Federation were to unite farmers nationally, and act as their pressure group in affecting government legislations. Small farmers, however, complained that the Federation was conservative and biased in favor of large commercial farmers in its policies. The National Farmers' Union was formed in 1960 with headquarters in Saskatoon, to serve as a loose federation of provincial unions which came into existence in Manitoba, Saskatchewan, Alberta, Ontario, and British Columbia between 1930 and 1955. The decreased farm incomes in the early 1960s due to drought, costly transportation, inadequate policy for feed grains distribution, and poor economic plight of small farmers increased the militancy of the National Farmers' Union, and in 1969 it was reorganized under a federal charter for the purpose of representing rank and file farmers regardless of the type of their agricultural activities. Their goals, policies, and orientations were so different that several attempts to bring the Canadian Federation of Agriculture and National Farmers' Union together were unsuccessful (Griezic, 1975: XVIII-XIX).

Canadian Federation of Agriculture

With the disintegration of the Canadian Council of Agriculture in 1932, there was no farmers' organizations left to provide leadership at the national level. Since the United Farmers of Canada was unable to fill this void, the organizational leadership passed to the commercial farmers' organization, and, as noted above, the Canadian Chamber of Agriculture was established in 1935 by the farmers' commodity groups and cooperative business enterprises. In 1939, the organization was renamed as the Canadian Federation of Agriculture (CFA).

The CFA represents farmers' organizations from all provinces. The CFA is a federated organization of its provincial counterparts. Each provincial unit is constituted of individual members as well as organizations at the district and county levels. Quebec, however, has three separate organizations which are independently affiliated with the national body. The Board of Directors of the CFA is made up of representatives from the provincial federations. The provincial federations are assessed membership fees based on the estimate of income of farmers in each province.

The CFA functions according to three major organizational principles: (1) political neutrality, (2) cooperation with government, and (3) a unified farmers' movement (Dawson, 1960). The political neutrality of the CFA implies that it sees the farmers' movement as an economic rather than a political one. It also enables the Federation to deal successfully with federal and provincial governments regardless of the parties which are in power. This relates to the second principle of the Federation of continuing attitudes of cooperation and negotiations with governments. A major goal of the CFA is to maintain a unified farmers' movement by coordinating the actions of various farmers' groups and associations into a cohesive national policy.

The Federation keeps the farmers informed of its long range aims, policies, and day-to-day activities through various channels of information and provincial federations. It concentrates most of its time in dealing with the government although it does not have a formal lobby group in Parliament. At both federal and provincial levels the Federation officials appear before legislative committees when matters which are of interest to farmers come up for discussion. The CFA submits an Annual Brief to the federal cabinet every year in which it summarizes its reactions to government policies. Aside from maintaining formal and informal contacts with the government departments, the Federation attempts to gain or increase membership on various administrative committees which enables it to put forward the farmers' view regarding the policies under discussion.

Dawson (1960) observes that the CFA has two major organizational weaknesses relating to financing and communication. Since farmers cannot individually obtain membership of the Federation, and since the CFA is a coordinating body of the provincial federations, the Federation and individual farmers are several steps away from each other which stands in the way of effective communication between the two. Policy decisions of the Federation flow from the top while financing flows from the bottom. This situation affects the effectiveness of the CFA as a spokesman for farmers and its own financial security: "At the very best it is fair to say that the Federation, with no direct membership, lacks contact with the farmers; at worst, it is possible many farmers do not even know that it exists."

National Farmers' Union

The alleged conservatism and commercial orientation of the CFA, and the desire to provide an effective voice for small farmers gave rise to a resurgence of farmer dissidence from 1939 to 1960. Several militant farmer organizations came into existence in the Prairie provinces between 1939 and 1947, and in Ontario and British Columbia in 1952 and 1955 respectively. An Interprovincial Farm Council was formed as an organizational alternative to the CFA in 1947. Their major areas of differences with the CFA were: "the method of affiliation with the Federation, the cooperatives' right to speak for rank and file farmers, and the means of organizing farming provincially and pursuing farmers' demands" (Griezic, 1975: XIX). The differences were so insurmountable that amalgamation discussion with the CFA broke down twice, in 1954 and in 1962.

The National Farmers' Union (NFU) was formed in 1960 as a loose federation of provincial groups with its headquarters located in Saskatoon. By the time it was reorganized under a federal charter in 1969, it had affiliated groups in almost all provinces. The NFU purports to represent small farmers of all types regardless of the nature of their agricultural pursuits. The major objects of the Union are: (1) legislation guaranteeing farmers' collective bargaining rights in the marketing of farm products; (2) legislation permitting the formation of producer controlled national marketing boards; (3) restrictions on corporate structures entering into the primary production of food; and (4) restrictions on nonfarm and foreign investment in the ownership of farm lands. The NFU is also in favor of maintaining family farms, and halting the process of rural depopulation and the disintegration of rural communities.

The National Farmers' Union is different from the CFA in structural terms as well. Individual farmers are considered members directly of the national body, not just of the local and regional groups. The Union is divided into several regions which may or may not correspond with provincial boundaries. For example, while the province of Ontario constitutes a Region by itself, the provinces of New Brunswick, Nova Scotia, and Prince Edward Island together constitute a single Region. Each Region is constituted of several Districts; and each District is made up of several Locals. The Board of Directors at each level is made up of representatives from the subordinate levels. The policy and decision-making ideally flow from the bottom to top, although this ideal cannot always be strictly conformed to.

While the CFA maintains continued relationships of cooperation with governments in attaining legislation beneficial to farmers' interests, the NFU takes a confrontational approach through strikes, pickets, tractor parades, boycotts, and free distribution or destruction of farmers' produce to express their grievances and demands. Neither of these farmers' organizations, however, endorses independent political action. The CFA is a much larger organization than the NFU.

The farmers' movement in Canada is thus not as intense at present as it had been in the first three decades of the twentieth century. Since the mid 1930s the movement has become more economic rather than political in orientation. The goal of the movement is to protect the economic interests of farmers rather than attaining political control as a social class. Some of the significant results of farmers' agitation in recent years are: greater government assistance to agriculture, legislation for increased credit and crop insurance, marketing boards, and programs to assist rural communities to adjust to social, economic, and technological changes. But such government measures have been piecemeal and have not been successful in securing higher returns for the farmer.

Social and Institutional Consequences

The technological transformation of agriculture has had far reaching consequences for rural social life. The decline in the number of farms caused by consolidation of holdings and capitalization of agriculture has contributed to extensive outmigration from farming and rural communities. The rural depopulation, in turn, has profoundly affected the community, family, school, religious and political structures of rural Canada. The extent and nature of these social and institutional changes are discussed in the two following chapters.

SUMMARY

While the technological revolution has raised agricultural productivity in phenomenal terms, its consequences for the rural and farming communities have not necessarily been beneficial. Agriculture has been industrialized to the point that very few of the activities traditionally associated with farming are conducted on the farm. Farming has been integrated with nonfarm industries giving rise to agrifood industry. While the rise of the agrifood industry epitomizes the modernization of agriculture, farming has become the occupation of a select few in rural societies. The capital-intensive nature of modern agriculture makes it impossible for the large majority of rural people to stay in or take up farming as an occupation. Not only are rural living and farming not synonymous any more, very few of those who remain in agriculture operate economically viable and profitable operations. A small number of successful farmers who partly own and partly rent very large, and highly mechanized and commercialized units, produce most of the food and fibre in Canada. Although the majority of these large and productive units are family-operated, they bear no resemblance with the traditional family farms. The remaining operators are either part-time farmers or operate medium to small farms of limited productivity, and depend on incomes derived from off-farm sources. The majority of the farms in Canada experience declining returns from farming due to a lack of needed resources and capital, an inability to use the most modern technology, difficult and often unsuccessful competition with large farmers and agrifood industries, and the cost-price squeeze.

Farmers in Canada have traditionally joined cooperatives and farmers' organizations during periods of social and economic distress to collectively promote and protect their interests. Agricultural cooperatives, especially marketing and supply cooperatives, have grown in number, membership, and volume of business in recent decades. Although these cooperatives have enabled many farmers to strengthen their bargaining and competitive positions in the market place, their emphasis on economic profit have failed to arouse the farmers as a class to gain some political control over the course of their destiny. The larger of the two existing national organizations of farmers, the Canadian Federation of Agriculture, eschews confrontation with government, and follows a cautious and pragmatic course in protecting farmers' interests. The National Farmers' Union, which is more militant in its activities in promoting the interests of farming communities, is much smaller in size and influence. Although both of these organizations have been reasonably successful in putting pressure on government to bring in legislation to promote the farming sector, the effects of these programs have not been significant enough to reverse the tide of the declining number of farms and financial return from farming. Rural depopulation and the decrease in the number of farms and farmers brought about by the modernization of agriculture also have had far reaching social and institutional consequences for the rural communities.

CHAPTER FIVE

Rural Social Structure: The Community and Family

The three preceding chapters dealt with the demographic and economic transformation of rural Canadian society. The technological change and the substitution of capital for labor in agriculture contributed to the rural outmigration and the decline of the occupational importance of farming in the rural economy. The rural depopulation and the rapid disappearance of traditional family farms, in turn, have profoundly affected the rural social structure. The changes in social structure, however, have not been caused solely by these demographic and economic forces. The diffusion of urban traits and values has also been an important factor in rural social change. Indeed, the effects of technological revolution in agriculture and urbanization on the rural social structure are so closely interrelated that they can be differentiated only in analytical terms. The changes in the traditional rural social life are most clearly evident in the rural community and family structures even though they continue to maintain features which are still distinctively rural.

THE RURAL COMMUNITY

The term "community" is used in many different ways by social scientists. Most sociologists, however, view the "community" as a territorial group consisting of people living together in a common territory and maintaining reciprocal relationships with each other which emerge spontaneously out of physical proximity and a consciousness of kind. Based on his analysis of 94 definitions, Hillery (1955: 111) found that there was a consensus among sociologists regarding three structural elements of the community: (1) a geographic area, (2) people in social interaction, and (3) one or more additional common ties. Hillery thus defined the community as consisting "of persons in social interaction within a geographic area and have one or more additional common ties."

The referent of "community" is represented by a great variety of territorial groups: neighborhood, village, small town, rural-urban fringe area, suburban community, small city and metropolis. As communities, they are all territorial groups, and are organized around patterns and activities that occur on a very regular or daily basis: shopping, schools, work, church, local government, some forms of recreation and so forth. But they are also vastly different in population size, density, and heterogeneity, the characteristics which, according to Wirth (1938), differentiate "urban" from "rural" communities. As they grow in population size, density, and heterogeneity, communities increasingly attain urban qualities. These community types thus can be successively arranged on the rural-urban continuum scale with the rural neighborhood being located near the "rural end" and the large metropolis near the "urban end" (Rogers and Burdge, 1972: 166-8). The rural urban fringe area stands half-way between the rural and urban poles. In this section we will focus our attention on the three types of rural communities — rural neighborhood, village, and town-country community — and the community in the rural-urban fringe.

The Rural Neighborhood

"Neighborhoods" are small in size, include one or a few simple services, and are characterized by an intimate social interaction among member-households. Neighborhoods may grow around a one-room school, a church, a grain elevator, a cross-road store or a combination of several of these. They may also grow in a mountain hollow, a cove, a river valley or may be formed by some other types of geographical barriers which mark them off from the larger community of which they are a part. This relative isolation from the larger community contributes to their distinctive physical and social identity which is often expressed in a neighborhood name, a strong sense of belonging among residents, and frequent "neighboring" including mutual visiting, exchange of favors and services, and joint participation in social and recreational activities.

The distinction between the "neighborhood" and the "community" is not very precise, and some sociologists prefer not to make any conceptual distinction between the two at all (Alexander, 1952: 237). However, some distinction can be made between them in terms of size, number and quality of available services, and intensity of relationships among residents.

A neighborhood is not only smaller in territorial terms but also in number of people. Because of the smallness of its size, relationships among members of a neighborhood are primary; and it provides only a few basic services to its residents. On the other hand, the community is larger both in geographical and population size, less intimate in social relationships, but self-sufficient in meeting most of the institutional and services needs of its residents. Thus, while the "community is a group of people within a localized area containing a majority of the major social institutions," the neighborhood is a localized group with "one or two social institutions as the focal point or means" by which it can be identified physically (Bertrand, 1958: 77).

Systematic studies of rural neighborhoods are almost nonexistent in Canada. Charles and Anthony (1943), however, used the term "neighborhood" in describing the small one-room school-centered communities in Prince Edward Island: "a group of people who live within easy reach of each other, who share a number of social activities, and who recognize collective responsibility for mutual aid." They identified several reasons for the close correspondence between the one-room school districts and neighborhood boundaries. Since the boundaries of many of these districts were determined by natural features such as rivers and marshlands, the neighborhoods were relatively isolated. The functioning of the school depended on close cooperation among the residents of the district. Finally, the pattern of obligation set by the responsibility of providing education to children extended into other spheres of neighborhood life as well. The community which thus developed within a school district served a variety of purposes, especially in providing a means of social security through mutual aid during times of distress.

Charles and Anthony referred to these entities as "neighborhoods" rather than "communities" because of their smallness in size, lack of institutional self-sufficiency, and close face-to-face relationships among residents which do not occur in a larger community. As neighborhoods they were parts of a larger community, a larger village or small town, which provided the services not available within the neighborhood boundaries.

The two neighborhoods studied by Charles and Anthony in the mid 1940s were small (147 and 91 persons) and each had a school, a church, a store, and a saw or grist mill. Mixed farming was the principal source of livelihood for at least two-thirds of the households in each. Other residents worked in the mills within the neighborhoods or outside. The school was the nucleus of neighborhood identity and interpersonal relations. The satisfactory functioning of the school depended on the close cooperation between the school trustees and the local Women's Institute. The church also played an important role in the neighborhood by raising funds for social purposes. The neighborhood residents, however, often travelled to the nearest town or surrounding neighborhoods to shop, go to movies or dances, and visit friends and relatives. Although local activities thus had overlapping geographical boundaries, the area over which a strong sense of solidarity was felt and intimate relationships maintained was clearly delineated by the neighborhood boundaries.

A more recent study of three such neighborhoods in Prince Edward Island reported considerable changes (Dasgupta, 1984). The physical isolation, one-room schools, and small farms that formed the foundation of these neighborhoods have now disappeared. The network of all weather roads and highways constructed in recent decades has left few communities isolated. The process of school consolidation which started in the early 1970s has led to almost total disappearance of one room school districts. Small-scale and mixed farming has been largely replaced by large, specialized, and capital-intensive operations. The occupational self-sufficiency maintained by these neighborhoods even as late as the mid 1940s is now a thing of the past. Only about one-third of the households in the three neighborhoods were of farmers or fishermen; those of the remaining two-thirds either survived on government pension and welfare or worked outside their neighborhoods. Not only did most residents work elsewhere to earn an income, the territorial circumference of their shopping and trading activities moved far beyond the horizons of their neighborhoods, a consequence of mass use of automobiles and centralization of services and institutions in the larger towns and cities.

The consolidation of the one-room schools and the loss of economic functions, however, have not necessarily meant the decline of the neighborhoods as "affective entities" defined in terms of social, recreational, and kinship ties and a strong sense of identity with the neighborhood. When asked to name the community they belonged to, for example, the respondents overwhelmingly mentioned the name of their neighborhoods.

Although in the past the one-room school provided a central focus for solidarity, rural neighborhoods in the Island were also characterized by other structural mechanisms which reinforced a common identity and shared consciousness: religious and recreational activities, visiting and mutual aid, and extended kinship. Church-going in all three neighborhoods of the study was almost totally confined within the neighborhood boundaries, and the churches continued to play an important role in the life of the neighborhood, also being used for recreation and social functions such as singing, dancing, and fund raising for various community purposes.

Mutual cooperation and visiting were also major structural features. Friends and especially relatives within the neighborhood itself were most frequently visited.

Mutual support among relatives who were also often neighbors was an important source of economic and emotional security.

Rural neighborhoods in Prince Edward Island thus appear to be strong and socially cohesive despite the loss of self-sufficiency and institutions such as the one-room school. The world of affective relations of islanders is still largely confined within their neighborhoods with neighbors, kin, and co-religionists often of the same ethnic background. This homogeneity constitutes an enduring structure enabling rural neighborhoods to persist as affective entities even though they have declined in economic terms.

The findings of these two studies of neighborhoods in Prince Edward Island are consistent with those of other studies conducted elsewhere in North America. Consolidation of schools and churches, centralization of institutions and services in larger centers, decline of small-scale agriculture, outmigration of residents to towns and cities for education and employment, and increased acceptance of urban ways of life are some of the factors which threaten the survival of rural neighborhoods. On the other hand, the "emotional unity" of the residents, strong identity with the neighborhood, and close ties based on kinship, neighborly activities, and religious and ethnic ties tend to help neighborhoods survive as strong affective entities in spite of the economic and functional decline caused by the forces of modernization and urbanization (Bertrand, 1958: 77).

The Trade Center: Village and Town

The rural village or small town in Canada is typically a "trade center" resided primarily by nonagriculturists: businessmen, shopkeepers, artisans, and professionals. The village or small town serves as a trade center to the farmers who live on the "scattered" farms widely separated from one another in the open country. While the "scattered settlement pattern" of North America has certain economic advantages in that agricultural activities are carried out in the immediate neighborhood of the home, and the location of the farmstead on the farm allows a highly integrated and coordinated economic activity, it also has certain obvious social disadvantages. The neighbors in such a settlement pattern are widely scattered over a large territory; and the lack of territorial proximity of neighbors and their institutions stands in the way of a strong feeling of community solidarity and identity comparable to that found in a nucleated agricultural village.

A variant settlement pattern, called the "rang" or "line settlement," is found in Quebec. In the line settlement, farms are long, narrow, and rectilinear with one end being located on a major transportation artery: a stream, a river, or a highway. The farmsteads thus line one or both sides of the waterfront or roadway. The closely located farmsteads constitute what is generally referred to as the "line village" or "string village." The settlement pattern of the rang allows farmers to build their houses on their large farms and at the same time the layout of their farms protects them from isolation. The proximity of the houses along the same road or riverbank, according to Deffontaines (1964: 12-14), gave rise to close neighborly relations in the past. Each rang had its own council, school, chapel, prayer hall, butter and cheese dairy and women's circle. A "unifying spirit of brotherhood pervaded the relationship of mutual help and cooperation among the neighbors in a rang." Garigue (1971: 127), however, contradicts Deffontaines in his description of the rang as a social unit: "the equal spacing of farms along a single straight line or *rang*, had the ... effect of preventing the rise of strong community feeling because of the great distance between the opposite ends." He maintains that even though rangs are characterized by a pattern of social relationship and reciprocity between neighbors, they never became a strongly integrated social unit because of a lack of central focus.

The villages and towns were established in Canada for the utilitarian purpose of performing services of communication, transportation, trade, and industry to farmers living in scattered settlements. They are centers of business rather than residence of farm families. Not all trade centers, however, depend on the surrounding rural area for their survival. Some trade centers are large enough in population size to survive on their own while others depend entirely or moderately on their rural hinterland for their existence. Based on his survey of trade centers across Canada, Laskin (1968) classified them into three major categories: (1) nonagricultural centers, (2) semiagricultural centers, and (3) agricultural centers.

"Nonagricultural centers" are represented by large towns or villages which do not specifically depend on surrounding agricultural areas, and are functionally self-sufficient. Such a center would not be seriously affected if the agricultural area around it ceases to exist. These towns have their own industries and commercial establishments, and although the farmers from the surrounding rural area take advantage of their services, their survival does not depend upon rural patronage.

"Semiagricultural centers" are towns which serve as trade centers in a partial sense. Although they have distinct agricultural service functions in addition to one or two other important economic interests, they do not rely heavily on labor supplied by the surrounding rural area nor would they be seriously affected if the agricultural area around them ceased to exist. In general, larger communities with a population of 5000 or over are likely to be nonagricultural or semiagricultural centers.

"Agricultural centers" are small villages or towns with populations of less than 5000. They primarily serve the surrounding rural area as a service center. Of the 327 towns and villages surveyed in agricultural areas by Laskin, over two-thirds were of the agricultural center type. Laskin categorized the agricultural centers into three subtypes based on their degree of dependence on the surrounding agricultural area: (A) "moderately" dependent; (B) "seriously" dependent; and (C) "entirely" dependent. The great majority (60 percent) of the agricultural centers were of the subtype B; and subtypes B and C together accounted for 90 percent of the agricultural centers surveyed. Both of these subtypes depended heavily on rural trade and drew significantly upon the surrounding rural area for their labor force. The rural children attended the schools; and the farmers attended the churches, and actively participated in the organizations located in these towns and villages.

The Town-Country Community

The "town-country community" is based on the interdependence between a trade center and the surrounding open country. It is "an area with a trade center where there is a network of relationships involving nearly all the common interests of the residents who recognize the community as a cohesive social unit to which they feel they belong" (Slocum, 1962: 386). Locating the boundaries of such a community is, however, a difficult task. The two major ways of delineating the boundaries of a town-country community are: (1) identification method and (2) neighborhood cluster method.

The "identification" method involves asking the residents of the open country which trade center they feel most closely identified with, and obtain certain services, send their children to school, and attend church, and then drawing lines around the area. This is very similar to the pioneering method adopted by Galpin (1915) who determined the boundaries of a town-country community by asking local merchants of a town or village to locate their most distant open country customers on a map. The trade zone of a rural village or town to Galpin was also the actual territory within which characteristic affective relationships of a community obtained. The economic emphasis implicit in such a technique has come under criticism in recent years (Haga and Folse, 1977). With the mass use of automobiles, the territory of trading activities has expanded in phenomenal terms but the area within which noneconomic relationships occur has remained confined to the limited geographical settings.

The "neighborhood cluster" method involves drawing up the map of a community by determining the area to which the people of a neighborhood feel most closely attached. This allows the investigator to attach a whole neighborhood, an already existing social unit, to the larger unit. Instead of asking the members of each individual household, key informants are asked to identify the trade center to which the members of the neighborhood in general feel attached.

Whatever the method, the boundaries of a town-country community cannot be precisely delineated for the members residing in the peripheral areas may identify with and patronize more than one trade center. Locating community boundaries has become especially difficult because the ruralite no longer patronizes and identifies with only one trade center but several at different levels starting from his own neighborhood center through a rural village, a small town, and a regional city although the degree and intensity of identification and patronage varies at different levels. While the relationships and participation at the lower levels tend to be frequent, personal, and affective, those at the upper levels tend to be less frequent, impersonal, and primarily economic.

Oyen: The Rural Village

Jean Burnet's (1951) description of the relationship of interdependence between Oyen, a rural village in eastern Alberta, and the surrounding open country in the late 1940s is an excellent illustration of the social organization of a traditional town-country community in Canada.

Oyen is located on the Goose Lake railway line, about 70 miles southeast of Hanna, the nearest city. The population of Oyen was 400 in 1946, and it was the locale of two important institutions: the only bank east of Hanna, and the Special Areas Office, a land controlling and seed distributing agency of the provincial government. Besides these two major institutions, Oyen also had general, hardware, and furniture stores, grain companies, implement dealers, a utility and power company, a flour mill, garages and filling stations, a drug store, a barber shop, restaurants, a hotel, a movie theater, a hospital, a high school, four churches, golf links, a curling rink, a lumber yard, and veterinary and dental clinics.

The open-country trading area of Oyen could be divided into an outer zone and an inner zone. The farmers of the outer zone chiefly came to Oyen for the bank, the Special Areas Office, the hospital, the implement dealers, the high school, the lumber yard, and the veterinary service. Since they did their ordinary shopping in their neighborhoods, the farmers of the outer zone were not intimately associated with the social and economic life of Oyen although their patronage was essential to the prosperity of the village.

The farmers of the inner zone visited Oyen four or five times a week not only to shop but also "to skate, play hockey, curl, play poker or bridge, knit, drink beer — in short, to act like villagers ... The young people actually live in town, though they return to the farm to sleep, eat, and work." The boundaries between the inner and outer zones, however, could not be determined in terms of distance from Oyen although it was one of the determining factors: "Some people six miles away belong to the outer orbit whereas some twenty-six miles away belong to the inner orbit... a map of Oyen's inner and outer zones would not look like a pair of concentric circles, but would show one zone scattered piecemeal throughout the other until the hinterland of another village center is reached."

There was little formal cooperation between the villagers and farmers but there was a tacit sharing of facilities, and a total lack of bickering between the two groups because a significant number of the villagers were also farmers who lived on their farms only during the working season. Although Oyen was an incorporated village, it was not set off from the rural community because the village council was not aggressive in asserting its distinctive identity. While only one of the four churches located in Oyen was primarily rural, the other three also had rural members not only in the congregation but also on their boards. Farmers participated in all the men's organizations in Oyen. Although they did not play golf, almost all curled, played poker, attended pool-halls and beer parlours, and belonged to various informal cliques. The young people from the surrounding rural areas were most attracted to the village. Oyen was their accepted meeting place on Saturday evenings, Sundays, holidays, and days of slack work. They regularly went there for dances or shows, for skating and hockey, or just for loafing. The young men, however, were in more often than young women probably because the young men enjoyed greater freedom, particularly in the matter of having a car.

In summary, "the country and village are between them able to support organizations of various kinds adequate to their needs, organizations which neither could maintain alone. The farmers are thus able to participate in a large number of activities which without an accessible village they would lack. The villagers form the core of the organization ... but the farmers attend as they can and as they wish. Both derive advantages and contribute essential support."

Oyen, as a relatively small agricultural village, represented a trade center which seriously depended on its trading zone for its survival. The participation of the farmers in the affairs of Oyen was not only economic but also social although such participation was not structurally formalized. The smallness and relative lack of urban orientation of Oyen contributed to a structural and cultural continuity between the village and the rural area. Oyen and its rural orbit thus constituted a harmonious and an interdependent social system.

The Small Town

The relationships characterizing those between Oyen and its hinterland do not necessarily obtain between a larger town and its surrounding area. In describing the relationships between Hanna, a town of 1400 people, and the rural residents, Burnet (1951: 75) noted: "Although the village of Oyen is a social center for the farmers who live near it, as well as a service station, the town of Hanna is only the latter. Hanna supplies more numerous and diverse goods and services than Oyen, but is not a center of rural life. It is regarded by the rural communities around as a distinct, and at times hostile, social entity."

As a railroad town, Hanna was able to provide a large variety of goods and services to both townsmen and villagers which also enabled it to be partially independent of the surrounding rural area. Although the economic independence of Hanna was more apparent than real, its large population allowed it some social freedom because the townspeople "could maintain among themselves churches, schools, clubs and lodges, without aid from either railroaders or farmers. The lack of social contacts, even enforced social contacts, increased the differentiation of rural and town ways of life."

Such a differentiation caused much friction and bitterness between the town and country. The farmers felt exploited by the town businessmen who appeared to control the supply of credit and local patronage because of their political influences. A genuine friendship and cooperation between the town and country failed to develop for several reasons: the ethnic heterogeneity of the rural population with the preponderance of German-speaking farmers increased the gulf between the country and the Anglo-Saxon townsmen; the constant growth of Hanna made it increasingly difficult for the farmers and townsmen to know one another; goods and services were increasingly being obtained by the farmers from the larger cities beyond Hanna by Mail-order or automobile trips; and there was little joint participation of farmers and townspeople in the vast number of associations of Hanna including churches, clubs, lodges, and athletic associations.

Hanna, the town, and its surrounding country were thus "economically interdependent but socially distinct." On the other hand, Oyen, the village, was a "part of the rural community and contributes to its well-being." Burnet rightly makes the point that although it is often claimed in the sociological literature that the distinction between a town and a village is obsolete, "the two terms are not altogether synonymous." The town has more nearly urban organizations and attitudes and usually a larger population, while the village is more intimately bound to the surrounding rural region and has a smaller population. The town thus tends to be a distinct social entity, whereas the village tends to be part of the functioning social system which includes as its integral part the farming region around.

Resurgence of Small Towns and Villages. Contrary to the popular belief, "Canadian towns and villages can no longer be considered a 'dying breed' of settlement" (Hodge and Qadeer, 1983: 215). There has been a "resurgence in the fortunes and rules of small centers" since 1961. There were over nine thousand small towns and villages with an average population size of 473 with a total of about 5 million people living in them in 1981. Between 1961 and 1981, the population living in such small centers grew by almost one million people. Thus, "small centers demonstrate that they are a significant and generally stable form of habitat in an otherwise highly urbanized, postindustrial nation."

Hodge and Qadeer (1983: 215-8) explain this staying-power of small centers in terms of three interrelated aspects: national, local functional, and community. Towns and villages persist by functioning within the "societal context" of contemporary Canada. The increasingly firm linkage of the people, businesses, and institutions of towns and villages to the larger society enables small centers to persist and grow. In the local functional context, a number of villages and towns located within an area of possibly several hundred square miles "collectively" provide the needed goods and services to the residents of a rural district without showing a great deal of dependence on a nearby city. The societal and the local (or subregional) contexts thus provide the "facilitating mechanisms," such as the physical and social infrastructures, necessary for small centers to persist. However, these mechanisms are not sufficient by themselves for the persistence of towns and villages. "Clearly, there must be an impetus from people and communities, acting in conjunction with the facilitating mechanisms of the two larger contexts, to allow a town or village to persist." In other words, the staying power of a small center depends very much on the community context as well, that is, "on the actions of people in the community — the individuals, groups, firms, and institutions."

The "Farm City"

The third category of trade centers are larger independent towns which came into existence in Canada after the Second World War, when good roads began to be built for year-round use, and trucks and automobiles enabled people and their goods to move over long distances in a shorter period of time. An expanded agricultural market during the war years, use of farm machinery powered by internal combustion engines, and the rising economic prosperity of farmers contributed to an increase in the market basket of the average rural family which even included bread and butter. The strategically located towns, especially those at the intersections of main highways and railroads, started to grow and prosper at the expense of smaller towns and villages and began to offer a large number and variety of services. The smaller towns and villages lost many of their economic functions, and came to be dominated by the larger towns. The growing independent towns became little cities which Zimmerman and Moneo designated as "farm cities" (1970: 6-28). Farm cities are distinct from larger cities in that fundamentally they are "elaborate retail and local administrative centers," rather than centers for wholesaling, manufacturing, and governmental administrative functions, and serve a large trading area which includes smaller towns, villages, and the open country.

Farm cities, however, provide only economic service to the hinterlands. The social activities and affective identity still remain bound within the geographic confines of small town and village-centered communities. While people still live in or around their small communities, they travel to these little cities three or four times a month where they get most of their elaborate services: banking, machinery, automobiles, medical, dental, and legal. But the basic human needs, "as for example schools, churches, cemeteries, friendships, and retirement homes" still come from smaller towns and villages. Although these smaller centers "of sympathetic, or Gemeinschaft, personal human relations" have lost some of their economic and leadership functions with the rise of farm cities, "contrary to most theories and ideas about the rural community, the minuscule organizations are not disappearing. ... Still they generally continue because they satisfy certain basic and intangible human needs."

The Rural-Urban Fringe

The urban society meets the rural at the rural-urban fringe. It is the buffer zone between the city and the countryside. The mass use of automobiles and other means of rapid transportation has made the growth of the fringe possible as more and more people from the city moved to the rural area just beyond its boundaries. A large variety of reasons motivate the people of the city to move to the country: to own a home paying an affordable price; to raise children away from the noise and congestion of the city; to live in or retire to a healthy environment with open space and fresh air; and to take up gardening or small-scale farming as a hobby. In his study of a small town in the rural-urban fringe called Paradise in southern Ontario, Barrett (1994: 99) found that the reasons often vary according to the class positions of migrants to the fringe. The majority of the newcomers belonged to the "middle, aspiring-middle, and lower-middle classes." Home ownership was the single most important reason for the newcomers who belonged to the last two categories: "Those in the lower-middle class were content to purchase modest dwellings, or to rent them, while the people in the aspiring-middle class often went up to their ears in debt in order to move into a dwelling compatible with their anticipated status." It was, however, not the lure of having a home of their own which made the newcomers of the middle or upper classes to move to Paradise. It was either the offer of an attractive job or the opportunity to establish a business of their own in or around Paradise was the principal motivating factor. The newcomers who belonged to the lower class simply "drifted into Paradise because of prior family connection."

There were several secondary reasons for the newcomers to decide to come to Paradise. One of them was to be able to afford financially to let one of the spouses, frequently the mother, stay home to look after the children. It was almost impossible to do so in the city with only one pay check. While they all enjoyed and appreciated the quality of life in the open and rural environment of Paradise, it was rarely the chief reason to move there. There were two other secondary reasons which were "unusual, if not bizarre." The first reason given by a large number of men was that they wanted to be able to have a large dog, frequently a German Shepherd, as a pet. The second reason which was also given by an equally large number of newcomers was that they "left the city in order to get away from the visible minorities."

Sim's (1988: 77-81) description of the fictional community of Ribbonville provides a composite portrait of a typical fringe community showing mixed rural and urban land uses, urban takeover of the countryside, increasing mutual interdependence of the city and the country, and the rise of a new kind of community with a distinct identity of its own.

A drive into the countryside beyond the outer limit of the city shows the urban intrusion in the land use pattern as evident in the "procession of motels, filling stations, fast food joints, contractors' assembly yards, and lumber yards, as well as tired looking houses with neglected front yards" which line the roadside not too far from a barn with a silo surrounded by farm machinery and cows. As one drives further farms become a regular feature some of which "have a house or two carved out of a front field, ranch-style structures right out of southern California." While one farm may have a "field full of bulldozers and backhoes," another with John Deere dealership may display "neat rows of green farm implements." Here stands an abandoned one-room school house "lovingly turned into a residence," there "stands a small white church surrounded by tall spruce trees" and a neatly manicured yard resplendent with a flower bed.

> Rural communities within commuting distance of cities are so common in all parts of Canada that a composite portrait is difficult to present. Moreover, the increasing intermingling of city and country elements call for very acute perception of the surviving rural aspect, which tends to be overshadowed by urban. But overshadowing does not necessarily lead to extinction. ...
>
> Providing bedrooms to commuters from nearby cities is not Ribbonville's only function. Commuting runs both ways. Not only are the roads filled night and morning with workers and professionals rushing to fill positions in the city, but there is a smaller reverse flow of city residents travelling to jobs in schools or small factories of Ribbonville. Daily commerce moves back and forth and crossways. ...
>
> The visitor driving along this paved ribbon fringed with nonfarm activities probably sees only the extension of the city into the country ... [and concludes] that nothing is going in the country that is not created and masterminded by powerful urban institutions. ...

> I do not intend to suggest that the attempted urban takeover of the countryside is exactly like an armed invasion. ... Nevertheless, invasion is an apt metaphor, for I have lived in rural communities that were criss-crossed by limited access highways, pipelines, and power corridors. These visitations of earth-moving equipment, macadam, and vibrating towers all came uninvited, were resisted, and were essentially accomplished by force. ...
>
> The urban invasion is not a simple thrust of economic and social forces from outside. ... The economic forces represented by business and allied government interests manifest themselves in the physical intrusions of pipelines, highways, and new housing. Rural people can master little defense against these forces. ...
>
> At the human level, the process is more complex. The old-timers may espouse commercial values and the newcomers environmental and human values. If there is a division it will be along ideological lines, with new and old residents, farm and nonfarm, lining up in strange new combinations. Out of these combinations the new culture in Ribbonville will grow (Sim, 1988: 78-81).

Sim is hopeful that the urban invasion of the countryside will not necessarily mean the decline of the rural qualities of fringe communities like Ribbonville. The fringe communities are not just new houses and gardens strung along roadsides like a ribbon in the open country, they also represent villages and towns "whose dwindling populations have been replenished and whose boundaries have been swollen by urban newcomers." These communities have a sense of fierce independence with distinct "traditions and antipathies, the local loyalties, attachments, and collective ambitions that animate and give meaning" to their existence and identity. These communities, Sim claims, "provide a locale and a region of gestation for a new rural culture." They represent "a new kind of countryside that is at once brash and tawdry, at once delightful and spacious." Sim concludes by stating that "while life goes on in Ribbonville, the city is near at hand, providing employment, excitement, and enrichment but its influence is only a shadow cast over living active organism, a new type of community struggling to establish a new identity."

The evolution of the rural-urban community through an intermingling of rural and urban, however, is not necessarily peaceful and devoid of conflict. As noted by Sim, such conflicts often occur "along ideological lines." Westhues and Sinclair (1974) provided a detailed account of such a conflict between the oldtimers and newcomers in a fringe community of about two thousand people in southern Ontario which they called by the pseudonym Fringetown, and provided an interesting interpretation of the conflict. Most oldtimers of the town sided with the reeve and his council over the construction of two twenty-five unit, three-storey apartment buildings at the edge of the village. The newcomers, who were "for the most part recent refugees from urban life" in Toronto, formed the Citizen's Action Committee (CAC) to fight against the construction of the apartment buildings and any development which they considered a threat against the preservation and beauty of the rural environment.

The conflict came to be defined not as between an action committee and a municipal government but as between newcomers and oldtimers. The newcomers claimed that they were not opposed to all forms of development but only to those which "could possibly become a burden on the uniqueness of this village." The oldtimers resented the interference of the newcomers in the development of the village: "These city people think that they know everything. They never had any say about what went on in Montreal or Toronto but here they think that they can just push us around." The newcomers were seen by the oldtimers to belong to the external society and thus had no right to criticize what went on in Fringetown.

The conflict in Fringetown took rancorous form involving name calling, shouting matches, and threats because, according to Westhues and Sinclair, "the two sides were not behaving in a common framework of political process" (1974: 84). While the newcomers felt that the reeve's handling of the municipal government was highly personalized and illegitimate, the reeve and the oldtimers thought that the activities of the CAC and newcomers fell outside the village's established means of political expression. The authors of the study suggested that the absence of party politics and party structures in the municipal affairs of Fringetown was highly conducive to the occurrence of rancorous conflicts because there was no institutional structure to inhibit them. The "structural strain" precipitated by a sizeable migration of newcomers challenging and threatening the traditional power structure of the village was also a contributing factor to the conflict. The final factor which played an important role in giving rise to the conflict was the absence of "structural integration" of the oldtimers and newcomers due to their differences in education, occupation, social class backgrounds and attitudinal patterns. It was interesting that it was the newcomers who were in favor of preserving an older way of life while oldtimers were in favor of modernization and development. The conflict in Fringetown was thus a conflict between two different ideas regarding the development of the village.

How does one distinguish between the "suburb" and the "rural-urban fringe"? Are they synonymous or conceptually distinct? The distinction is not always obvious or easy to make. Suburbs are generally built around the outer edge of the central city. A suburb is politically independent from the central city but economically dependent upon it. The rural-urban fringe, however, is an area beyond the established suburbs of a city showing convergence of urban and rural characteristics. While the suburb exists outside the political boundary of a city but within its zone of influence, the rural-urban fringe may or may not be under the direct influence of a metropolitan city.

The Emerging Rural Community in Canada: An Overview

The traditional rural community in Canada was organized around neighborhoods, villages, and small towns. During the era preceding mass use of automobiles, the small trading area of a village or town served as the territory within which residents interacted both for economic and social purposes. The inhabitants of a town-country community felt a strong sense of identity with it, and cooperated both in economic and social spheres of activities. The farmer practised subsistence agriculture and produced almost all that the family needed as well as some surplus to trade in the nearest village or town for cash. Since the bulk of farmers lived in "rangs" and neighborhoods, this is where most of their daily economic and social activities took place. Economic cooperation in the form of exchange of labor, implements, and services, and social interaction manifested in frequent visiting, neighborly cooperation, and joint participation in recreational activities constituted the basis of social life. The nearest village or town, however, "was the nucleus of the community" (Fortin, 1971: 92), although a trip to the village, located sometimes at a distance of three or four miles or more, was not every frequent for those families who did not have access to the needed means of transportation. Farmers travelled to the village to trade, shop, attend the church, and for a variety of other services. The village thus bound together several neighborhoods around it into an identifiable community. In Quebec, for example, the village, where the church was located, and a number of neighboring "rangs" constituted not only the boundaries of a parish but also of a rural community: "Because a church attracted the population of the scattered rangs to a central point, it also became the best location of other activities. Shops opened in its vicinity, and craftsmen, professionals and others came to live close to it ... It permitted the gradual self-perception by the people living close to it that they had common interests which differed from the interest of other groups" (Garigue, 1971: 138).

Several factors contributed to the decline of the rural community as independent, self-sufficient, and relatively isolated socio-cultural entities: commercialization of agriculture, rural outmigration, improved means of transportation and communication, and centralization of services and institutions in larger towns and cities. Although the initiation of the process predated it, World War II helped in its accentuation. Farming became profitable with great increase in demand for agricultural goods for export. In many rural community agriculture for the first time became an economic activity involving cash returns. Profits earned from farming were now invested in modern tools and techniques, and in buying consumer goods such as radios and automobiles. The commercialization of agriculture forced farmers to increase the size of farming operation to maintain a profitable output or quit farming and move to urban centers for jobs in industries. The attraction of a better income and the lure of a better life also caused the rural youth to migrate to the urban and industrial centers.

A considerable proportion of farmers who stayed behind in rural communities earned a higher income, spent more on consumer goods, and enjoyed a better standard of living than before. Since it was impossible for small rural communities to satisfy the growing needs of rural residents, both institutions and services came to be centralized in towns and small cities. Better roads and use of automobiles enabled rural residents to travel farther and farther from their communities to obtain specialized goods and services to satisfy their increasingly urbanized needs. The greater contact with urban centers meant the "breakdown of old traditions and the diffusion of urban traits into the country" (Miner, 1971: 64).

The increased mobility of rural residents has given rise to at least two new levels of community integration beyond that represented by a village or a small town and its surrounding rural area. The first is represented by an independent town and its hinterland which includes several dependent small towns and rural villages. Since the independent town is larger, it maintains a degree of social self-sufficiency although it depends to some extent on trade with the people of its hinterland. Farmers and rural residents obtain goods and services from the town but have very little involvement in its affairs, and seldom identify with it. Such a town, as exemplified by Hanna, maintains an identity distinct from its rural hinterland. The relationship between the town and country, at this level, is mostly "functional" or economic.

The integration between a rural area and a regional or "farm" city is totally functional. A farm city is large enough to be the locale of a great variety of institutions and services, and serves a much larger territory which includes both independent and small towns, and rural villages. The relation between a farm city and its hinterland is economic, formal, and impersonal.

While the relations at these two levels of community integration are largely functional, those at the lower levels are predominantly "affective" or noneconomic in nature. Although the small rural community has lost many of its economic functions, it continues to survive as an affective entity satisfying social and emotional needs of its residents. Studies have shown that the territory of community identity and close social interaction did not expand concomitantly with that of functional activities. While "essential activities such as work, education, and shopping, tend to occur in a much larger region ... people continue to relate symbolically to limited geographical settings. Insofar as community is based on sustained face-to-face interaction, the regional setting should seem unlikely to foster it" (Wilkinson, 1978: 118).

Rural residents thus relate to a series of succeedingly wider levels of community integration each serving a distinctive function for them. The rural neighborhood made up of a few closely interacting households is more a community of symbolic identity and affective relationships rather than an entity satisfying the functional needs of its residents. The composite town-country community made up of several rural neighborhoods integrated into a community around a rural village or a small town continues to survive as a social unit, and satisfies the basic economic, social, and emotional needs of its residents even though it has lost its relative functional self-sufficiency and autonomy. The community which has evolved around a large independent town includes a much larger territory, and is primarily formal and functional in integration. While the independent town offers a wide variety of services and institutions to the population within its hinterland, the townsmen and the country people remain socially distinct. The regional farm city serves even a larger territory offering a greater variety of services and institutions; but it functions primarily as a center for economic activity, and its relationship with the rural hinterland is almost entirely gesellschaft.

Hodge and Qadeer (1983) note the emergence of yet another kind of rural community in recent decades which has been designated as the "dispersed city." They maintain that the rural community structure made up of a series of succeedingly wider levels of community integration — the neighborhood, village, town, and farm city — each serving an increasingly larger hinterland with greater and more specialized varieties of goods and services has undergone a significant transformation. Hodge and Qadeer point out that although today's villages and small towns may not look very different from Burnet's description of them in the 1940s, their roles in the rural society have shown important changes: "Though continuing to be places of concerted activities (and establishments), small communities are ceasing to be singular focal points for surrounding populations." The lateral links between a village or town and the surrounding hinterland have weakened as rural residents obtain their goods and services from a variety of trade centers.

Based on his observation of a similar phenomenon in the United States, Hart (1975: 168) described the new spatial organization as the "dispersed city." The dispersed city includes several trade centers each offering one or two specialized goods and services and operating as elements of an interlinked system of places. It is thus neither an actual city nor an urban entity but a form of spatial organization of villages and towns which as trade centers with specialized functions collectively provide some of the goods and services normally found in a city. Just as the city dweller travels to other parts of the city in search of goods and services, so does the modern resident of a town or village who drives to the other towns and villages within the territory of the "dispersed city" to obtain the goods and services he or she needs.

Fred Dahms (1996) looked for empirical evidence of the existence of the dispersed city in the Wellington County of southern Ontario. The Wellington County is situated mid-way between Toronto and London. He found that although many of the villages and towns in the county declined both in population and as trade centers with increased use of automobiles, a number of them continued to survive and prosper by providing specialized goods and services and offering local employment to residents and commuters: "Although these settlements are separated by productive farmland, the mobility and flexibility afforded by the car have enabled some functional unity to develop among them." It is possible to shop, work and live in spatially separate settlements within Wellington County linked by their mobile inhabitants. Access to specialized goods and services provided by the towns and villages within the county is almost as often and convention as access to work or shopping for residents of a congested metropolis. Many residents of the Wellington County live and shop in the towns and villages "rather than consistently travelling to higher order centers for major shopping."

THE RURAL FAMILY

The farm family constituted the foundation of traditional rural communities. In describing the farm families in a small community in rural Quebec in the 1930s, Gerin (1971: 49) made the following observation.

> Almost all of them were engaged in farming, an activity which is performed in common by their own members, paid workers being hired only in exceptional circumstances; they complement farm resources with other natural productions of the land or the sea, or with revenues from handicrafts and various accessory trades. These families will thus be all equal, or at least, they will not form a hierarchy. Anything that a family cannot do or produce through the cooperation of its own members will be done or made available as far as possible by neighbours, on an exchange basis. This was the situation as observed by Gauldrec-Boilleu in 1861, and the circumstances have not changed significantly since that time.

The rural family in Quebec continued to manifest the same traditional features until the Second World War. This was evident in the control of the family over its members. The strongly familistic orientation of the rural family fused its members into an integrated organic unity with close mutual affection and regard for each other, and enabled it to function as a relatively stable entity over generations.

The rural family in Canada shared many of the features of a traditional rural family described above until the changes brought about by the technological revolution in agriculture and the rapid diffusion of urban traits. This is evident in the description provided by Fortin (1971: 94-6) of the family in an agricultural parish of Quebec. The family "constituted the social unit *par excellence*. Due to its comparative isolation, even within the *rang*, the family was the real unit of subsistence and it was supposed to meet all the needs of the individual. Its functions were therefore very numerous."

The economic function of the family was to produce all the needed goods through the joint participation among its members in the productive activities. Even the children who worked away from the farm gave their wages to the father to spend or invest for the family or farm. The values attached to family and agriculture were transmitted to the children by their parents under whose direction and supervision they learnt to be farmers or housewives gradually through their participation in all farming and household activities. In other words, "it was the family much more than the school which acted as the training centre for the young."

The family was also a unit for recreational and religious activities. Most of the leisure time of the members was spent either in the family or in the informal social gatherings within the neighborhood. Family prayer was also the most important form of religious activity, for it was often difficult, especially during the winter months, to go regularly to the church located several miles away in the village.

Although the traditional French and English-Canadian rural families differed in certain respects, they were very similar in basic functional features (Elkin, 1964: 33):

> This rural family was, above all, a relatively independent economic unit, serving both as a unit of production and consumption. The husband, wife, and children had clearly defined tasks ... The father was the leader of the household but the wife often exercised a major influence in family decisions.
>
> Larger kinship ties were taken for granted. Everyone felt that he could count on relatives and neighbors to help in an emergency and in turn he would go to great lengths to give similar services. Birth rates were high and children were valued as workers on the farm as well as in their own right.
>
> Religion was crucial in giving integration and cohesion ... Family life, in its daily tasks and in the ceremonies of life cycle, was imbued with the sacredness of a religious spirit ...
>
> Thus, in this simplified and somewhat idealized picture of a rural society, we find closely knit families, with relatively little change from one generation to another, with sex and generation statuses clearly differentiated, and with the family's pattern of life and thought sustained by the surrounding larger family, community and church.

It was the joint participation in the subsistent activity on the family farm and its relative isolation from the society at large which constituted the foundation of the traditional rural family. The common interest and a sense of cooperative effort developed by the participation of its members under the patriarchal authority vested in the father helped to knit the rural family together as a self-sufficient entity with an attitude of independence. Its relative isolation not only forced it to produce its own food, clothing, and other requirements but also made the rural family independent of urban and industrial centers. The traditional rural family in Canada, in other words, represented a structural adaptation to a socioeconomic environment characterized by small scale subsistence farming done with relatively primitive technology on abundantly available arable land in isolated communities with which it maintained a degree of equilibrium (Fortin, 1971: 96). As the environment started to change, the equilibrium was lost, and the attainment of a new equilibrium became necessary.

Rural Family in Transition

The rural family which adapted to a life of subsistence farming in a relatively isolated self-sufficient community has changed significantly with the increasing commercialization of agriculture, and contact with urban and industrial centers. These changes are evident in the rural family's number, size, and composition as well as in its internal role relationships and social and economic functions.

Decline in Number

The number of rural families has consistently declined both in absolute and relative terms over the decades until 1981, when for the first time it showed an increase over that of the preceding decade (Table 5.1). It was, however, the rural nonfarm families which increased in number in 1981 while the rural farm families continued to decline. Urban families also continued to increase in number although they showed a slight proportional decrease in 1981. The proportions in 1991 were almost identical to those in 1981. Although the farm families declined both in number and percent from 1981, the nonfarm families compensated for this decline by showing an increase both in absolute and proportional terms. The number of urban families continued to grow although they showed a negligible proportional increase over that of 1981.

The consistent increase in the number of families in urban areas where almost four-fifths of the country's population is located, is a reflection of the fact that the total number of families in Canada is rapidly growing with the rising marriage rate.

Almost two-thirds (63 percent) of Canada's adult population was married in 1991 compared to only 50 percent in 1901 and 56 percent in 1931 (Statistics Canada, 1979: 2-3). Although the increase could be partially attributed to the fact that there are now more persons in the marriageable age group (20-34) than ever before, an increasing proportion of men and women feel financially secure enough to get married, and often at an earlier age.

TABLE 5.1

RURAL AND URBAN FAMILIES IN CANADA, 1961-1991

Area	1961		1971		1981		1991	
	No.	Pct.	No.	Pct.	No.	Pct.	No.	Pct.
Canada	4,147,444	100	5,070,681	100	6,324,975	100	7,356,170	100
Rural	1,162,389	28.0	1,147,305	22.6	1,518,180	24.0	1,760,185	24.0
Farm	428,808	10.3	305,155	6.0	251,685	4.0	214,665	3.0
Nonfarm	733,581	17.7	842,150	16.6	1,266,315	20.0	1,545,520	21.0
Urban	2,985,055	72.0	3,923,380	77.4	4,806,795	76.0	5,595,980	76.0

Source: Census of Canada

The rural-urban difference in the number of families is accounted for by several factors. First, the size of the rural population in Canada is much smaller than the urban. Second, the rural population has a significantly smaller number of marriageable men and women because of the rate of outmigration among the rural youth. Finally, the lack of financial solvency compels the rural youth to marry at a lower rate and at a later age than the urban youth.

Size of the Rural Family

Rural families have traditionally been larger than urban families. Subsistence and labor intensive agriculture, and economic importance of children for the contributions they make to farming activity are two major factors associated with the high value attached to large families in traditional rural societies. A large family ensures its continuity over generations and a steady supply of labor for the family farm. Children are also considered a source of economic support and security in old age. Urban people, on the other hand, tend to prefer fewer children in the family for it to be able to maintain a higher standard of living.

TABLE 5.2

SIZE OF RURAL AND URBAN FAMILIES IN CANADA, 1991

Area	Total number of Families	Number of Persons in Family			
		Two	Three	Four	Five or More
Canada					
Number	7,356,170	3,139,725	1,671,735	1,722,620	822,090
Percent	100	42.7	22.7	23.4	11.2
Rural					
Number	1,760,185	696,470	377,240	427,645	258,825
Percent	100	39.6	21.4	24.3	14.7
Farm					
Number	214,665	73,410	43,000	50,080	48,175
Percent	100	34.2	20.0	23.3	22.5
Nonfarm					
Number	1,545,520	623,065	334,235	377,765	210,655
Percent	100	40.3	21.6	24.5	13.6
Urban					
Number	4,450,445	1,904,390	1,039,460	1,049,750	456,840
Percent	100	42.8	23.3	23.6	10.3

Source: Census of Canada

Although the rural-urban gap in the size of the family has been declining in Canada, some difference continues to persist. The proportion of two-person families, for example, is higher in urban than in rural areas in Canada (Table 5.2). The incidence of two-person families is lowest in the rural farm. On the other hand, while the rural farm areas have the highest percentage of families with five or more persons, the urban areas have the lowest. Three or four-person families, however, occur more or less in similar proportions in farm, nonfarm, and urban areas.

Decrease in Size

The average size of family in Canada has been declining in recent decades. The trend of decreasing average size can be observed both for urban and rural families although rural families continue to be larger than their urban counterpart (Table 5.3).

Among rural families, the average size of farm families is larger than that of nonfarm families.

TABLE 5.3

AVERAGE NUMBER OF PERSONS IN RURAL AND URBAN FAMILIES IN CANADA, 1961-1991

Area	Average Number of Persons in Family			
	1961	1971	1981	1991
Canada	3.9	3.7	3.3	3.1
Rural	4.3	4.1	3.5	3.2
Farm	4.5	4.3	3.8	3.5
Nonfarm	4.2	4.0	3.4	3.2
Urban	3.7	3.6	3.2	3.0

Source: Census of Canada

Mechanization of agriculture, an increasing standard of living and education, and diffusion of urban values related to family living to rural areas appear to have contributed to the decline of the average size of the rural family. The replacement of labor by capital in agriculture has removed the need for a large family labor on the farm and raising a large family. The availability of social insurance and pension benefits provide economic insurance for old age. The modernization of agriculture has raised the overall standard of living in rural areas although the economic disparity within the farming population is higher than ever. The increasing availability of educational resources and a change in the attitudes of the rural people toward formal education have raised the level of educational attainment in rural areas. Finally, the improvement in mass communication and transportation and increased mobility of the rural population have contributed to the diffusion of urban oriented values of family living to rural areas. The rural family is thus not only getting smaller in size but also more nuclearized in composition.

Increase in Nuclearization

A traditional rural family often includes members of two or more related nuclear units, generally that of the father and of his son or sons living together in a joint household. This intergenerational nature of the rural family enables it to maintain a large group of people working together on a common farm. The strong emphasis on kinship ties and obligations acts as an integrating force in keeping the family together both as a social and economic unit, and allows it to maintain self-sufficiency and independence.

Members of an extended family, however, do not necessarily live together in a joint household. The nuclear units of an extended family are often divided into separate but spatially proximate households, and continue to maintain close cooperation and fulfill mutual obligations. Such extended units could be referred to as "segmentary extended families" as opposed to the "joint extended families" comprising two or more related nuclear families living together in the same household. Indeed, in rural India, segmentary extended families are more common than joint (Dasgupta, 1993).

In Canada there is a general emphasis on the nuclear family made up of husband and wife with or without unmarried children, and the solidarity of the married couple is stressed to the exclusion of brothers, sisters, and parents. This is true for both urban and rural families which were overwhelmingly nuclear in that over four-fifths were husband-wife families in 1991 (Table 5.4). Indeed, the proportions of husband-wife families were higher in rural farm and rural nonfarm than in urban areas. This was primarily due to the fact that a significantly larger proportion of urban families were lone parent families. On the other hand, the incidence of lone parent families was lowest in the rural farm population. The relatively conservative tradition of the rural farm population is further evident in the significantly smaller proportion of common-law couples among its husband-wife families compared to those of rural nonfarm and urban.

TABLE 5.4

RURAL AND URBAN FAMILY AND HOUSEHOLD TYPES IN CANADA, 1991

Family and Household Types	Canada		Rural		Farm	
	No.	Pct.	No.	Pct.	No.	Pct.
Family Type						
Total Families	7,356,170	100	1,760,185	100	214,665	100
Husband-Wife Families	6,401,460	87.0	1,608,425	91.4	204,510	95.3
Married Couple	5,675,510	77.2	1,440,680	81.8	197,065	91.8
Common-Law	725,950	9.9	167,745	9.5	7,445	3.5
Lone-Parent Families	954,710	13.0	151,765	8.6	10,155	4.7
Male Parent	168,240	2.3	38,895	2.2	4,370	2.0
Female Parent	786,470	10.7	112,870	6.4	5,785	2.7
Household Type						
Total Households	10,018,265	100	2,129,365	100	236,115	100
Non-Family	2,763,035	27.6	391,935	18.4	23,655	10.0
One Family	7,118,655	71.1	1,715,245	80.6	210,330	89.1
Two or More Families	116,570	1.3	22,185	1.0	2,135	0.9

Family and Household Types	Non-Farm		Urban	
Family Type				
Total Families	1,545,520	100	5,595,980	100
Husband-Wife Families	1,403,915	90.8	4,793,040	85.7
Married Couple	1,243,615	80.5	4,234,830	75.7
Common-Law	160,300	10.4	558,210	10.0
Lone-Parent Families	141,610	9.2	802,940	14.3
Male Parent	34,530	2.2	129,345	2.3
Female Parent	107,080	6.9	673,605	12.0
Household Type				
Total Households	1,893,250	100	7,888,900	100
Non-Family	368,280	19.5	2,391,100	30.3
One Family	1,504,910	79.5	5,403,410	68.5
Two or More Families	20,055	1.0	94,390	1.2

Source: Census of Canada

While families in Canada have generally been nuclear in composition, a proportion of households with two or more, often related, families have always been in existence. The proportional incidence of such multi-family households, however, has always been low in Canada and in recent decades has almost become insignificant. The proportions of multi-family households have been around one percent in rural farm, nonfarm, and urban populations in 1991 although, contrary to expectation, they occur in the highest proportion in urban and lowest in rural farm. The higher incidence of multi-family households in urban areas can be explained in terms of lack of affordable and the relatively high cost of housing in cities. The same factors also explain the significantly higher incidence of nonfamily households in urban areas than in rural farm and nonfarm.

The households in Canada sometimes also include one or more additional persons related (unmarried, widowed, separated, divorced or elderly kin) or unrelated with the family or families they comprise. Sixty percent of such nonfamily persons lived with the families of their relatives or nonrelatives in 1991 in rural farm areas, whereas the corresponding percentages for rural nonfarm and urban were 47 and 44. In other words, the farming families included one or more related or unrelated persons in a significantly higher proportion than the nonfarm and urban families. A higher proportion of these additional persons were related to the families they lived with in rural farm and nonfarm areas while the proportions were opposite in urban areas.

Over the last several decades families in Canada have become more nuclear, a result of urbanization. This trend is evident not only in urban but also in rural families. Whatever economic encouragement there was for the formation of joint families, especially in the form of "stem families" due to the prevailing system of primogeniture, it disappeared with the commercialization of farming. Verdon (1980: 105-24), for example, noted that the Quebec stem family as it came into existence during the pioneer era was an economic arrangement between the father and one of his sons who lived together in a joint household with their families until the father died and the son became the sole heir of the total farm. The son who showed continuing interest in agriculture, and was prepared to accept a subordinate role to his father in the stem family, was chosen as the heir of the farm which he inherited only after his father's death. It was, in other words, "the reluctant covenant of two nuclear families, with opposed interests, to co-reside for purely economic reasons." But transformations in agriculture changed the developmental cycle of the residential group in rural Quebec. Sons started to play key roles in the management of farms because of their greater literacy and innovative skill as farming became increasingly specialized and market-oriented. Since fewer sons wanted to stay on their paternal land due to increasing competitiveness in farming and opportunities for better income off the farm, "farmers almost have to lure one of their sons into taking over." The mode of inheritance has changed and fathers now either "create corporations with their heirs or sell their estate to the heir in exchange for an annual allowance." The result is the disappearance of the stem family. Although close neighbors, "the father's and the heir's families now form separate residential groups."

The near-absence of joint households in rural communities, however, should not necessarily be considered as indicative of a complete absence of extended family or kinship. Extended families, as noted above, do not always manifest themselves in joint units. Instead of sharing the same household, related families in rural areas often live together in close proximity in the same neighborhood and maintain close relations of mutual obligation and cooperation.

Extended Kinship

The rural family in Canada continues to maintain its identity in its larger size and emphasis on extended kin ties even though like its urban counterpart it is predominantly nuclear in household composition. Indeed, the rural farm family, the most nuclear of all, is the largest in average size. This is perhaps a continuation of the heritage since frontier times when large but nuclear households constituted the basic social and economic units in rural Canada. Although the immediate family is generally the most important unit from the standpoint of residence, making a living, and consumption, the tendency for the rural family is to include a wider scope of kin in activities beyond the boundaries of the immediate family. This is expressed not only in their greater concern with lines of descent than urban families but also in a feeling of greater sense of responsibility for unattached and dependent kin. The emphasis on kin ties originates in the fact that rural people tend to marry more within their neighborhood and community, and consequently are related with most of their neighbors by blood or marriage.

In their study of a farming and ranching community in the Jasper region of Saskatchewan, Kohl and Bennett (1974: 209) noted that within the community "there were numerous affinal connections, and households were connected through more than one affinal tie." The family and kinship ties created a strong community sentiment to the extent that it kept the sons of ranchers and farmers of the region from moving out of the community. While nuclear households were idealized, and the majority were nuclear units, the practice of one of the married sons living in the parental household to take over the ranching or farming enterprise after the retirement or death of the father created "a *de facto* form of the extended family" in many instances (similar to what Verdon described as "stem family" in rural Quebec). The practice enabled the sons of a rancher or a farmer to gradually buy in installments or inherit their father's enterprise. If it was not large enough for several sons to succeed, the father either built up his enterprise or bought additional land to permit the support of more than one family. The father moved to the nearest town after retirement, "and since Jasper roads are difficult to maintain in severe weather, the town grandparent emerges as a key figure for the school-age child. The grandparental house offers the child a place to stay in town overnight. During severe winters it is common for the grandchildren to spend the school week at the grandparents' home in town and come home to the ranch on weekends." Contacts with kin living outside the community or region were maintained with regularity. They visited the farm or ranch on their way to east or west or made it their vacation destination. "Thus ties are established between the young generation and their extended kin out of the region, and hospitality is extended by the rural family so that at a future time when they call on their urban relatives for reciprocity it can be accepted freely."

Several studies in Prince Edward Island have also shown extended kinship to be an important structural element of the rural communities. Of the two neighborhoods studied by Charles and Anthony (1943: 47-8), all households in one and almost two-thirds in another were linked by kinship or marriage. They observed that "the ideals of social obligation are first extended to relatives ... and secondly to all neighbors." In another study (MacDonald, 1960: 78) it was found that "a large number of families in the community were related by kinship and marriage and shared far more in common activities than those who were not related." A similar pattern was found to exist in three communities on the Island studied more recently (Dasgupta, 1982: 261-5). The families which exchanged visits frequently and maintained relations of mutual obligation were also often related by blood or marriage. The family structure was typically that of a nuclear family with close extended kin ties in the community. The mutual support among kinsmen was an important source of economic and social security. Extended kin ties along with ethnic and religious homogeneity constituted the major structural principles in terms of which rural communities in Prince Edward Island maintained their integration and identity.

Studies, however, also indicate that although the rural people continue to be more kin-oriented than their urban counterpart, the emphasis on kinship in rural communities has declined with urban influence (Reimer, 1997: 90). The authors of the Sterling County study in Nova Scotia (Hughes et al., 1960: 203) noted in describing the family and extended kinship in Fairhaven, a Protestant-English community, "major features such as size and connection with relatives have been reduced from what they were years ago; but it still is a strongly functioning unit, and the extended family bonds ... are stronger than they are in most urban centers." A comparison with Lavellee, an Acadian community in the same county, however, revealed that extended family bonds were weaker in Fairhaven. But even in Lavallee there was a weakening of extended kin ties and sentiments compared to what was in existence in the past.

The persistence of kinship recognition among French Canadians even in the face of urbanization has been explained by Garigue in cultural rather than structural terms. Based on his findings of persistence of kinship among French Canadians even in an urban environment, he argued that emphasis on kinship does not necessarily decline with urbanization; on the contrary, "kinship and urbanism among French Canadians seem to have become functionally interrelated" (1956: 1098). Garigue contended that the decline in kinship recognition as found among English Canadians is a function of cultural values rather than urbanization. Rioux (1971), another French Canadian sociologist, offered the counter argument that kinship is a part of the social structure rather than of cultural values, and that cultural values themselves become transformed to adapt to structural changes caused by urbanization. Since urbanization in rural French Canadian society has not proceeded to the extent it has in English Canada, extended family and kinship ties continue to function among the rural French Canadians. Kinship recognition and associated cultural values in rural French Canada will be transformed with further urbanization as they have in English Canada and the United States. Whyte (1968: 55) seems to concur with Rioux by stating that while the family was the cornerstone of the relatively isolated early French Canadian social structure, "the English Canadians never became as closed and enveloped, due largely to its ties with British and American institutions." Therefore, while French and English rural Canadian families are similar in functional terms, "structurally the French Canadian rural family is more impervious to the inroads of urbanization."

Internal Relationships

Structurally, the rural family has become smaller in size and more nuclear in composition with a decreased emphasis on extended family and kinship ties although it continues to manifest greater recognition of kinship than the urban family. The increasing structural resemblance with the urban family is also reflected in the rural family's changing internal role relationships.

A study of internal relationships in rural Saskatchewan (Royal Commission on Agriculture and Rural Life, Report 10, 1956: 62) revealed that the emerging rural family manifests a combination of traditional and modern forms. The division of labor is no longer maintained in strictly age and sex lines. With the increasing modernization of agriculture, and greater emphasis on formal education in rural societies, children's role in farm work has greatly diminished. The tradition of putting children to work around the farm at an early age is almost nonexistent. The increasing school and social activities of older children make them less available for farm work, and those who do help their parents in farming activities no longer confine themselves to tasks traditionally differentiated according to sex. It is not uncommon for older girls, for example, to work with machinery and herds which were traditionally considered to be "boys'" or "men's work."

The Saskatchewan study also revealed a decline of the strong dominance the father enjoyed in the traditional farm family as a decision maker. Although he continued to play a major decision making role in buying farm machinery, in planting crops, and in financial affairs such as deciding to borrow money and allocating income among family members, the father shared responsibility with the mother and even with children for many other types of decisions.

A similar trend was observed in a study of Ontario farm families (Abell, 1975). Farm women no longer confined themselves solely to housework and child rearing. Nearly all of the 352 farm wives (97 percent) interviewed reported doing a variety of farm tasks traditionally considered to be "men's work" such as operating farm machinery, driving a truck or tractor, and keeping farm accounts. Increasing use of machinery both for domestic and agricultural purposes, and dependence on off-farm work to supplement income from farming are two major factors responsible for the changing role of farm wives. The use of modern household equipment has freed the rural housewife from the drudgery of housework, and enabled her to get involved in both on and off farm activities. As more and more farmers become gainfully employed off the farm to supplement their farm income, farm wives take increasing responsibilities in farming activities. Abell, for example, found in her study of Ontario farms that in a few cases where husbands took full-time employment off the farm, the wives "had taken over the entire responsibility of managing the farm, including the hiring of occasional labour for certain peak work periods." Availability of nonfarm work within commuting distance from the farm, and farm wives' ability to continue farming operation even in the absence of male operators thus make part-time farming possible.

The technological transformation of agriculture has thus altered the traditional husband-wife relationships on the farm. Although the husband continues to be the sole decision maker in certain areas, much of the daily planning and decision making in domestic and farming activities are shared by the husband and wife. For example, in one-half of the 352 families of Abell's study, the farm operator alone made decisions relating to the farm but among one-third, farm plans were made jointly by husband and wife. In making plans and decisions relating to the home, husband and wife participated together in one-half of the families, while in one-third of the families they were the responsibilities of the wife alone. Abell's findings thus suggest that as the farm wife invaded the man's work world in addition to performing her traditional domestic duties, the husband also moved into the arena of decision making in matters relating to home.

In a more recent study in four Saskatchewan communities in 1987, Smith (1992: 162-6) found that farm wives participated in decision-making jointly or on an equal basis with their husbands in day-to-day farming as well as in major farm and household acquisitions. Three out of every four farm women also performed some specific types of work on the farm frequently which included acquiring supplies, keeping books, and tending livestock. A fewer proportion drove farming machinery, delivered meals to the field, and even repaired farm equipment. Although the household tasks of preparing meals, doing dishes, washing and ironing clothes, and cleaning house were almost exclusively performed by the housewives, some husbands assisted their wives in the performance of such household tasks. Husbands were especially helpful to their wives in households with children by contributing a significant amount of their time in caring for children and helping them with their homework. Although there is indeed a tendency for "sex-specialization" in the tasks performed, farm wives have become an integral part of Canadian agriculture by participating in a variety of activities on the farm as partners of their husbands. Husbands, on the other hand, are showing an increasing tendency to participate jointly with their wives in making decisions on household acquisitions and to assist their wives in performing various household tasks.

The democratizing trend in the rural family is evident not only in the relationships between the husband and wife but also in those between parents and children. The children in the traditional rural family were raised under the strict authoritarian discipline of their parents. The older boys worked on the farm under the supervision of the father, and older girls and children assisted their mother at home. They were not compensated by any regular amount of money or wage although they received substantial aid from their parents to get started in farming. The Saskatchewan study (1956: 85) describes the change in the pattern of relationships as follows:

> The pattern of father-centered decision making becomes more democratic. There is less control over the activities of individual family members ...
>
> ... areas of decision are increasingly shared with the children, who thus may view parents as guides and advisors rather than patriarchs, even if benevolent patriarchs, who make all the decisions and whose wishes must never be challenged.

The relaxation of authority and strictness in child rearing was also noted by the authors of the Fairhaven study (Hughes *et al.*, 1960: 214). Childhood restrictions and demands were moderately low in Fairhaven. Although parents offered few rewards to their children, privileges were not withdrawn as a form of punishment. "Praise, reasoning, and pointing out models for good behavior" were the preferred methods. "But at the same time, the clarity and intensity of sentiments about the proper way to raise children and manage a family have diminished, and are subject now to more uncertainty, anxiety, and discussion."

In summary, as the patriarchal features of the rural family declines, family interaction increasingly occurs on the basis of equality among members. There is a greater equality in the statuses of husband and wife, and of parents and children, within the contemporary rural family. Despite these changes the rural family has not become exactly like its urban counterpart. It still manifests in its structural features many elements which characterized the traditional rural family.

Changing Family Functions

The structural transformation of the rural family is largely a consequence of its changing functions. The traditional context within which the large, and sometimes extended, rural family used to function has changed with the declining labor input, and increased use of machinery in agriculture which made dependence on large unpaid family labor unnecessary. The decline in the size of the rural family is thus attributable not only to the urban influence and its rising standard of living but also to the functional irrelevance of largeness. The enhanced mobility of the rural family caused by the rural to urban migration, and loss of its functions in supporting and caring for the aged or otherwise incapacitated members and relatives to other institutions have greatly contributed to its increased nuclearization, and decreased emphasis on extended family and kin ties. Finally, the changing pattern of intrafamily roles brought about by rural family's increased dependence on off-farm work, rise in part-time farming, and consequent change in its economic circumstances have contributed to the decline of the rural family's patriarchal feature.

The economic functions of the farm family have been greatly affected by the commercialization of agricultural activities. Most farm families now produce for the market rather than for their own consumption. The farm family now consumes a greater variety of economic and domestic goods and services than it produces. The differentiation of productive activities in agriculture has greatly reduced the scope of economic activities on the farm. The farm family no longer manufactures or repairs its own tools or equipment, produces its own seeds, feeds, or fertilizers, or markets its agricultural products. The domestic needs of the modern farm family are so numerous and varied that they cannot be satisfied from its own internal resources. It not only buys the items which it does not and cannot produce such as the automobile, refrigerator, cooking range, radio or television but also the items such as butter, cheese, bread, wine, home canned goods, and sewn clothings which it could produce. The commercialization of farming activity and increasing involvement of the farm operator or the spouse in a full or part-time off-farm occupation enables the farm family to earn enough cash to purchase rather than produce the goods needed for domestic consumption.

The transformation of the economic function of the rural family can also be seen in the fact that the vast majority of the rural families are nonfarm, earning their income from occupational activities in service, commercial, transportation and communication, construction, and manufacturing and processing industries. Dissociated from farming, rural families have assumed many functional features similar to urban families in both productive and consuming activities.

The rural family has changed not only as an economic but also as a social unit. Many of its social functions have either changed or have been taken over by other institutions. Traditionally, the rural family provided economic security and protection to its members and relatives in old age or in times of illness or unemployment. Many of these protective functions are now performed by the state or public agencies. Most of the religious functions in earlier days were performed within the family. The rural family was a religious group, and family prayers, saying of grace, Bible reading, and hymn-singing were important religious activities. These religious activities have now been largely removed from the home.

It was the family much more than the school which acted as the training center for the young who learned to be farm operators or housewives under the supervision of their father or mother. Traditional farming required little formal education, and the socialization and training obtained from home were sufficient for the young to grow up to be full-fledged members of the traditional society. The modern rural family has lost most of these educational functions as more and more rural youth enter schools, colleges, and technical and vocational institutions to prepare themselves for a much more technologically oriented agriculture or occupations off the farm. Formal education has increasingly become the norm in rural societies.

The rural family is losing out to formal agencies and institutions in the sphere of recreational functions as well. In earlier days leisure and recreational activities were largely performed within the family or the neighborhood as neighbors frequently visited each other to exchange the news of the day, enjoy a sing-song or play games. These family-centered recreations have now largely been lost to church, school, the community at large, and commercialized recreation.

The "defunctionalization," however, does not mean disorganization of the rural family. It continues to perform a number of very important tasks: reproduction, care of children and their primary socialization, satisfying of emotional and affective needs of members, and conferring status to them. Reproduction and early socialization of members continue to be the prerogative of both rural and urban families. The task of satisfying emotional and affective needs of members by the rural family has become increasingly important as alternative sources of emotional gratification in the rural setting are decreasing in number and significance, and the element of depersonalization which accompanies technological advance increases. Its role of conferring status to its members in society has similarly become more important as other structures for status ascription to individuals, such as neighborhoods and communities, decline in importance. Under conditions of increasing urbanization and declining importance of such primary groups as neighborhoods, and small local communities, the rural family remains the only surviving link between the individual and the larger society.

Urbanization of the Rural Family

It is obvious from the foregoing discussion that the rural family in Canada is increasingly becoming similar to the urban family. Compared to the traditional, the contemporary rural family is more numerous but smaller in size, more nuclear in composition, weaker in recognition of extended family and kinship ties, less patriarchal and more equalitarian in internal role relationships, and narrower in the scope of its functions.

It is, however, not the urban influences alone but changes of the rural society in demographic and economic characteristics as well contributed to the transformation of the rural family which brought it closer to its urban counterpart. As its occupational focus has shifted away from farming, the rural family is no longer able to maintain its solidarity based on the common occupation of all members on the farm. This increasing occupational heterogeneity provides fewer opportunities to the members of a rural family to be involved together in a common task even for recreation and festivities.

The absence of frequently shared activities, however, does not necessarily mean that the contemporary rural families lack solidarity or integration. Whyte (1968: 61-2) maintains that as "role specialization is increasingly associated with functional specificity rather than traditional and biological ascription," the rural family manifests "the transition from a kind of cohesiveness based on affective identification with members of a group, referred to by Durkheim as mechanical solidarity, to a form of integration based on rational interdependence, which Durkheim called organic solidarity." It is somewhat paradoxical, Whyte points out, that as the rural family is losing its affective cohesiveness it is also increasingly becoming more important as a source of emotional gratification to its members.

Despite the changes described above, the rural family maintains "certain forms and a minimal intensity of interaction, which are almost guaranteed by rural living no matter how great the influence of urbanization. These can serve as a foundation for personal relationships and integration" (Smith and Zopf, 1970: 60). Since work and family life are less easily separated in rural areas than in cities, and farming, even in its commercialized form, requires a certain amount of intense interaction and communality, the rural family continues to maintain a minimal degree of affective cohesiveness which enables it to satisfy the emotional needs of its members. Thus, "although relationships among members may be weakened by the reduction in the number of functions performed by the family, they also may be strengthened as a greater share of time and effort devoted to the smaller number of functions remaining" (Smith and Zopf, 1970: 61). Indeed, the constellation of functions in the rural family, "even that most affected by urbanization," continue to be greater than those in the urban, which gives the rural family greater coherence and unity, and often a greater degree of affective interaction.

SUMMARY

The commercialization of agriculture, rural depopulation, and urban influences have had far reaching influence on the rural community and family in Canada. The small rural community has lost both its economic and social autonomy, and has become integrated with a series of succeedingly wider levels of community organizations in economic or functional terms. Better roads and use of automobiles enable rural residents to travel far away from their residential communities in larger centers where goods and services of greater varieties and specialization have been increasingly centralized. The horizon of economic activity of the rural residents thus extends far beyond the boundaries of residential communities. The territory of community identity and intimate social interaction, however, has not expanded concomitantly with that of functional activities. The small rural community continues to survive as an affective entity satisfying the social and emotional needs of its residents.

The forces which transformed the traditional rural community organization have changed the rural family as well. The rural family is increasingly attaining the features which characterize its urban counterpart. It is smaller in size, nuclear in composition, less patriarchal and more equalitarian in internal role relationships, and narrower in its scope of functions. Despite these changes, the rural family still maintains a distinctive pattern in personal relationships, integration, and intensity of interaction which are almost guaranteed by rural living no matter how great the influence of urbanization.

CHAPTER SIX

Rural Social Structure: The Church, School and Local Government

The family, church, and school together constituted the institutional foundation of the traditional rural community in Canada. The family, as the "social unit *par excellence*," met numerous needs of the individual within the home. The church and school integrated the families together into a solidary community. The church, as a place of religious worship, brought the families of the community together to participate collectively in religious rituals, ceremonies and festivities to protect and promote their common interests. The community as a whole acted as a congregation with its traditions and mores regulating the rituals and ceremonies, and the rural clergyman held a position of authority and influence not only in the religious but also in the secular sphere. The "religious culture" of the traditional rural community provided a moral reinforcement to agriculture as a way of life and to the norms and values associated with rural living.

The role of the school in the traditional rural community extended far beyond the realm of education. It provided a structural locus for the members of the community to participate in activities together in dealing with matters involving the interest of the entire community. It was not only a place to educate children but also a community center for social and recreational purposes. The church and school buildings represented the community in symbolic terms.

The local government in Canada, however, did not come into existence until about a hundred years ago. It was in the towns, rather than in rural areas, where local self-governments were first established. Since their establishment, however, local governments have become an integral part of the rural social structure in many parts of Canada. It is at this level, the rural people come in direct contact with formal political institutions, and have a direct contribution to their functioning.

The forces which transformed the rural community and family, have also affected the rural church, school, and local government. The purpose of this chapter is to describe the transformation of these three aspects of rural social structure both in terms of their sources and consequences.

THE RURAL CHURCH AND RELIGION

Religious values, subsistence agriculture, and farm family, were the three interrelated elements which were of fundamental importance to the traditional rural community in Canada. Religious values emphasized the virtue of agriculture as an occupation and as a way of life, and reinforced the norms of relations in the farm family in moral and ethical terms. Subsistence agriculture and the associated parsimony in consumption pattern reflected the moral and ethical norms of the traditional community, and provided the economic focus around which the farm family was organized. The farm family, in its turn, conformed closely to the norms and values associated with subsistence agriculture and the rural way of life.

The commercialization of agriculture and increasing diffusion of urban values, especially since the Second World War, however, have had a secularizing influence on the traditional rural life in Canada. Fortin (1971: 96-7) presents a graphic description of this process in analyzing social change in Ste. Julienne, an agricultural parish in Quebec. Commercialization of agriculture and increased contact with urban centers altered the value orientation of the farmers, and disrupted the traditional equilibrium of Ste. Julienne based on subsistence agriculture, farm family, and religion. The religious values attached to subsistence farming and the farm family were replaced by the secularized norms associated with consumer orientation and aspiration to attain a higher standard of material living. The farm family, consequently, "ceased to be a production unit" by becoming "almost exclusively, a consumption unit," and the role of the father shrunk to that of a procreator and breadwinner rather than a leader with authority to manage productive activities. Thus the religious culture and a rural way of life which manifested themselves in farming activities and relationships within the farm family were largely replaced by a materialistic and secular culture of urban origin.

The available statistical evidence also show the decline of religious influence in rural Canada. Whyte (1968: 81), analyzing the responses to the Gallup Poll question on whether the influence of religion on Canadian life is increasing or decreasing asked in 1954 and 1962, found that a growing proportion of both urban and rural people throughout Canada believed that the influence of religion on their lives was decreasing (1954 and 1962 figures were: 27.6 and 32.9 percent urban; 23.6 and 42.5 percent rural nonfarm; and 0.0 and 39.9 percent rural farm). Whyte was especially impressed by the fact that while none of the rural farm respondents in 1954 perceived such a decline, almost 40 percent did so in 1962. Noting that this phenomenon was especially marked in such less urbanized regions as Atlantic Canada and Quebec, Whyte observed (1968: 81):

> the continuing process of rationalization which has been accelerating in Canadian cities during the last decade has only recently begun in rural areas. The fact that this change coincides with, or follows closely behind, the demographic and technological changes taking place in rural life, and that it is most pronounced in those regions where the impact of technology has been most recently experienced, strongly supports the hypothesis that the influence of organized religion will be further restricted as urbanization proceeds.

Indeed, by 1968 over two-thirds of all Canadians, compared to only one-third in 1962, believed that the influence of religion on Canadian life was decreasing. Since

1968 (Gallup Polls, 1979), however, the proportions of those who perceived the decrease of religious influence has declined somewhat (69 percent in 1976, 61 percent in 1977 and 53 percent in 1979), and those who perceived an increase in religious influence has gone up (18 percent in 1976, 22 percent in 1977, and 24 percent in 1979). In other words, by the end of the Seventies over one-half of all Canadians continued to perceive the decline of religious influence in their lives, and only one-fourth perceived that the opposite was true. The process of secularization has thus made deep inroads not only into urban but also into rural Canada.

Even though the process of secularization has affected all Canadians, some rural-urban differences continue to persist. The perception that religious influence is declining is stronger in urban than in rural Canada. Within rural Canada, the more urbanized segments of the population, such as rural nonfarm as opposed to the farm, and the more urbanized regions continue to manifest a higher level of secularization. Based on his analysis of the responses to Gallup Poll questions, Whyte (1968: 83-4) noted that a greater proportion of rural farm and nonfarm respondents than urban believed in the literal acceptance of the New Testament, in a fiery hell, and in a life after death; regularly said grace at meals; and regularly attended church or synagogue. An examination of the response pattern by the regions showed that while the respondents from the less urbanized regions of Atlantic Canada, Quebec, and the Prairie provinces were "strong adherents to orthodox religious views," those of the more urbanized regions of Ontario and British Columbia were "less literal in their interpretation of the gospel and less diligent in their practice of religion." There was thus a general correspondence between the extent of urbanization of a region and secularization of religious beliefs and practices.

The Rural Church: Characteristics and Transformations

The change in the religious orientation of the rural people has inevitably contributed to the transformation of the rural church both in structural and functional terms. This is evident in the changing patterns of church affiliation, church participation, and influence of the church.

Religious Affiliation

Canada is overwhelmingly a Christian nation. In 1991, four out of every five persons were either Catholic, Protestant, or belonged to an Eastern Orthodox group (Table 6.1). Roman Catholics constituted the largest religious group and together with Ukrainian and Polish Catholics they accounted for close to one-half of the total Canadian population. Only a little over one-third of the Canadian population were Protestant. Protestants in Canada are divided into a large number of subgroups. Some of the larger Protestant groups are: United Church, Anglican, Presbyterian, Lutheran, Baptist, Pentecostal, Mennonite, Jehovah's Witness, and Salvation Army.

TABLE 6.1

RELIGIOUS AFFILIATION IN CANADA, 1981-1991

Religious Groups	1981		1991	
	Number	Percent	Number	Percent
Catholic	11,402,605	47.3	12,335,255	45.7
Roman Catholic	11,210,390	46.5	12,203,620	45.2
Ukrainian and Polish Catholic	192,215	0.8	131,635	0.5
Protestant	9,914,580	41.2	9,780,710	36.2
Eastern Orthodox	361,560	1.5	387,395	1.4
Eastern Non-Christian	305,890	1.3	747,455	2.8
Jewish	296,425	1.2	318,070	1.2
Other	18,910	0.1	38,970	0.1
No Religion	1,783,530	7.4	3,386,365	12.6
Total	24,083,500	100	26,994,045	100

Source: Census of Canada

Catholics, however, have not always been the largest religious group in Canada. Indeed, Protestants clearly outnumbered them from the later part of the eighteenth through the first few decades of the twentieth century but the proportions of Protestants relative to Catholics have been declining consistently since then (Mori, 1987: 13). The proportion of Protestants fell below that of Catholics for the first time in 1971 and the trend continued through 1981 and 1991 (Table 6.1). The drop in the percentage of population affiliated with the major Protestant denominations — such as United Church, Anglican, Presbyterian, Lutheran, and Baptist — was primarily responsible for the proportional decline of Protestants. The major Protestant denominations accounted for the religious affiliation of more than one-half of the total Canadian population before World War II but it declined to only a little over one-fourth in 1991 (McVey and Kalbach, 1995: 363).

The third largest group, after Catholics and Protestants, representing almost 13 percent of the total Canadian population in 1991, had no religious affiliation at all. This trend toward increasing secularization is further indicated by the fact that the proportion of people with "no religion" has increased dramatically over the last three decades. From a minuscule proportion of less than one percent in 1961, the proportions of Canadians indicating no religious preference increased to 4.3 percent in 1971, 7.4 percent in 1981, and 12.6 percent in 1991. One, however, has to be cautious in concluding from this increasing trend of no religious preference that there is a growing absence of religious faith among Canadians. McVey and Kalbach (1995: 364-5) suggest that the large increase in the immigrants of Eastern non-Christian religions since 1961 might have been a contributing factor other than those that have been generally attributed to the gradual secularization of Canada. Since the proportion indicating no religious preference has been relatively high among south and east Asian immigrants, it is possible that they interpreted the intent of the question asked regarding religious affiliation not to mean "an absence of faith so much as an indication that respondents might not be presently active in a church, or specifically associating with any one church at the time." On the other hand, the introduction of self-enumeration procedure since the 1971 census in urban areas might have made it easier for the respondents to give an honest answer to the question in the absence of an enumerator rather than providing a socially acceptable response.

Protestants enjoyed a slight majority over Catholics in Rural Canada in 1981 even though they were significantly outnumbered by the latter in the urban population. The declining tend of Protestants in Canada was reflected for the first time in rural population as well in 1991 (Table 6.2). Although they continued to maintain a significant numerical dominance in the rural farm population, their proportion declined considerably in the rural nonfarm in 1991. Catholics continued to be the largest religious group within the nonfarm population. Roman Catholics generally tend to be represented in a greater proportion in urban areas, especially in those with a population of 100,000 or more, than most of the other religious groups. Ukrainian Catholics, on the other hand, are found primarily in rural areas. Among the Protestant groups, Presbyterians and Anglicans are more urban than other groups, although all Protestant groups, except the Hutterites and Dukhobors, have a greater proportion of their members in urban rather than in rural areas. Anglicans, Baptists, Lutherans, and Mennonites are the major Protestant groups among both farm and nonfarm populations of rural Canada.

TABLE 6.2

RELIGIOUS AFFILIATION (IN PERCENT) IN RURAL, RURAL FARM, RURAL NONFARM, AND URBAN POPULATIONS, 1981-1991

Religious Groups	1981				1991			
	Rural	Farm	Nonfarm	Urban	Rural	Farm	Nonfarm	Urban
Catholic	44.9	35.9	46.8	47.7	44.8	33.4	46.4	46.0
Protestant	47.8	55.6	45.8	39.3	43.5	55.6	41.7	34.0
Other*	1.1	1.8	0.9	5.0	1.3	1.8	1.3	6.8
No Religion	6.2	5.2	6.5	8.0	10.4	9.2	10.6	13.2
Total	100	100	100	100	100	100	100	100

* Includes people belonging to Eastern Orthodox, Eastern Non-Christian, Jewish and other groups.
Source: Census of Canada

Religious groups other than Catholics and Protestants — which principally include the Eastern Orthodox, Eastern non-Christian, and Jewish — account for only a little over one percent of the rural Canadian population. They are mostly found in urban areas: "typically, the 'native' religions are disproportionately represented in the rural areas, whereas the 'foreign' religions, those of more recent introduction into a society, are located predominantly in urban areas" (Smith and Zopf, 1970: 351).

The proportion of rural people with no religious preference was only slightly less (10.4 percent) than their urban counterparts (13.2 percent) in 1991. The increase in the proportion of such people in 1991 from that in 1981 was almost as high in rural as in urban areas. Even among the rural farm people, who are typically conservative and religious, over nine percent of the respondents claimed to have no religion. The manifestations of this extreme secularization of beliefs and values thus presents itself across the rural farm-urban continuum although it is likely that a proportion of the rural farm and nonfarm people with no religious preference was made up of those who moved recently from urban to rural areas.

Church Membership and Attendance

Church Membership. Although over four-fifths of Canadians indicated their affiliation with a religious group, they did not all formally belong to a congregation. It is important to note that membership to a congregation is not necessarily stressed in all denominations or religions. Members of many non-Christian groups in Canada do not consider regular attendance in congregational service or rituals is of primary importance to the true believers. Even a number of Protestant groups, especially those of the Pentecostal and nondenominational persuasion, do not emphasize formal church membership. Among Catholics and most Protestant denominations, however, the tie between identification with the faith and membership of a congregation is strongly stressed. In Canada, with an overwhelmingly Christian population identifying with Catholic or Protestant faith, the trend in church membership is a good indicator of secularization.

In his study of the declining influence of religion, Bibby (1993: 3-11) describes the "continuous exodus" from church membership in Canada. Between 1957 and 1990, for example, Canada suffered a dramatic decline of 23 percent in congregational membership, from 52 to only 29. Over two-thirds of Roman Catholics in Canada stopped participating in congregational activities during the period. While outside Quebec congregational membership among Roman Catholics fell from 91 to 38 percent, within Quebec the decline was from 99 percent to only 18. Although not as large, Protestant denominations were not too far behind Roman Catholics in their 43 percent decline in church membership: from 80 percent in 1957 to 37 percent in 1990. Three major denominations — United Church, Anglican, and Lutheran — suffered a membership slide from more than 80 percent to around 30 percent. Even the more conservative denominations such as Baptist and other evangelical groups which boasted of a 75 percent membership in local congregations as recently as 1985, declined to only 55 percent in 1990.

Besides denominational variation, church membership in the past showed regional variations as well with the more rural and agricultural regions, such as the Atlantic and Prairie provinces and Quebec, showing a higher rate of church membership than the more industrialized and urbanized provinces like Ontario and British Columbia. Bibby's analysis (1993: 9) of the 1957 data on church membership showed the continuation of a similar pattern with some interesting exceptions. Quebec led the country in church membership (97 percent) followed by the Prairie provinces (82 percent), and British Columbia came last (54 percent); but the Atlantic Provinces, the most rural of all regions, tied with British Columbia, while Ontario, the most industrialized and urbanized of all regions, came third (70 percent) after Quebec and the Prairie provinces. In 1990, with the exception of Quebec which declined from 97 percent to only 16 percent in church membership, the proportions in other regions, although showed significant diminutions from 1957, followed the pattern of higher church membership for more rural than for more urban regions: Atlantic provinces (49 percent), Prairie provinces (36 percent), Ontario (31 percent), and British Columbia (27 percent). The pattern of decreasing rate of church membership with increasing urbanization was noted in the variation by community size as well. Although communities of all sizes suffered significant declines in congregational membership between 1957 and 1990, those with a population below 10,000 continued to have the highest percentage of church membership (33 percent), followed successively by those with a population size of 10,000-99,000 (31 percent), and more than 100,000 (27 percent).

The rate of congregational membership also showed variation by sex, age, and education both in 1957 and 1990. Women, for example, had a slightly higher rate of church membership than men. People belonging to the oldest age group 50 and over were members of a congregation in the highest proportion (40 percent) in 1990 while the rate of church membership among people belonging to the youngest age group 18-29 was the lowest (17 percent); and those of the 30-49 age group were found in between the oldest and youngest age groups in the rate of their membership of a congregation (26 percent). There was an increasing rate of church membership with increasing level of educational attainment: people with at least a college degree in 1990 showed the highest rate (41 percent) followed successively by those with some post-secondary education (25 percent), and a high school education or less (24 percent).

Regardless of the pattern of variation — whether by denominations, regions, community size, sex, age, and education — the comparison of the 1975 and 1990 clearly shows that an increasingly low level of church membership characterizes a majority of Canadians: "in all parts of the country; in small communities as well as the large cities; who are either women or men; of all educational levels; [and] who are older as well as younger" (Bibby, 1993: 10).

Church Attendance. Church attendance rate also declined as significantly as that of church membership. There was, according to the Gallup Poll, a drop of 24 percentage points during the period between 1965 and 1998. The question addressed to a representative sample of Canadians was: "Did you yourself happen to go to church (or synagogue) in the last seven days?" The percentage who responded affirmatively was 55 in 1965, 44 in 1970, 35 in 1980, 32 in 1985, 27 in 1990, and 31 in 1995. The attendance rate went down by one point in 1996 (30 percent), by another three points in 1997 (27) but rose by four percentage points in 1998 (31 percent). In other words, since the 1980s the weekly church attendance rate in Canada levelled off in the thirties although it came down twice, in 1990 and 1997, to high twenties. Working with the General Social Survey data of Statistics Canada on church attendance, Bari and Mori (1991: 23) reported a three percent drop in weekly church attendance, from 26.9 to 24.0, during the period between 1985 and 1990 (Table 6.3). Over the same five-year period there was a similar three percent decline among adults who attended church at least once a month but less than once a week. There was, however, an increase of 8.2 percent — from 18.9 to 27.1 — of adults who went to church at least once a year but less than once a month.

TABLE 6.3

CHURCH ATTENDANCE OF POPULATION AGED 15 AND OVER IN CANADA (IN PERCENT), 1985-1986, 1988-1990

Frequency of Attendance	1985	1986	1988	1989	1990
At Least Once a Week	26.9	27.2	27.3	25.9	24.0
At Least Once a Month	15.3	14.9	13.5	11.9	12.3
At Least Once a Year	18.9	20.3	26.3	26.7	27.1
Less Than Once a Year or Never	27.7	25.7	20.5	20.2	22.8
Not Stated/Not Applicable	11.2	11.9	12.4	15.2	13.8
Total	100	100	100	100	100

Source: Bari and Mori, 1991: 23.

The decline in the church participation rate among Roman Catholics in Canada has been phenomenal in recent decades. Between 1965 and 1998, according to the Gallup Poll, attendance among Roman Catholics fell by 47 percent, from 83 to only 36. During the same 33 year period the participation rate among Protestants showed a two percent increase from 32 to 34. This is probably explained by the fact that the process of secularization affected Protestants much earlier than Roman Catholics as indicated by the attendance figures of 1965: 83 percent among Roman Catholics and 32 percent among Protestants. It is only since the mid-1970s that the church attendance among Roman Catholics started to fall in a significant proportion with the French-speaking Catholics experiencing the severest drop (Mol, 1976: 251). Based on the analysis of his Project Canada survey data conducted every five years since 1975, Bibby (1995: 124) observed that as of the mid-1990s the weekly church attendance was considerably lower among Roman Catholics in Quebec (24 percent) than among Roman Catholics outside of Quebec (38 percent) suggesting the continuation of the trend of a higher rate of decline in church attendance among French-speaking than among English-speaking Roman Catholics first observed in the mid-1970s.

Church attendance shows a similar regional variation as has been observed earlier for church membership: the more rural and agricultural regions of the country tend to have a higher rate of church attendance than the more industrial and urban regions. For example, Bibby's (1993: 9-11) data on church attendance show that in 1957 the rate of weekly church visit was highest in Quebec (83 percent) followed by the Atlantic (50 percent) and Prairie (48 percent) provinces whereas in the more provinces of Ontario (44 percent) and British Columbia (25 percent) the church attendance rates were lower. Although the church attendance rate fell dramatically in all regions by 1990, the pattern of regional variation remained more or less the same with the exception that Quebec which led all regions in weekly church attendance in 1957 had one of the lowest attendance rates in 1990. According to the Gallup Poll data of 1998, the weekly church attendance rates were still higher in the Atlantic (33 percent) and Prairie (31 percent) provinces than Quebec (26 percent) and British Columbia (26 percent) but, interestingly enough, the rate was highest in Ontario (35 percent), the most urban and industrial province in the country.

The trend of higher church attendance rates in predominantly rural than urban areas is also borne out by the data on the community size and weekly church attendance. Bibby's (1995: 127) 1975 data showed declining percentages of church attendance for communities of increasing population size: 39 percent for communities with less than 10,000 people; 33 percent for those with 10,000 to 99,000; 27 percent for 100,000 to 400,000; and 26 percent for communities with over 400,000 people. In 1995, although communities with less than 10,000 people continued to have the highest church-going rate, those within the other three population size categories had almost identical percentages of church attendance (21, 19, and 22). Bibby pointed out that while the inverse relationship between the size of the community and church attendance changed very little for Protestants during the two decades between 1975 and 1995, with their strongest church attendance rate being concentrated in towns and smaller cities, the attendance rate among Catholics was down in communities of every size. There was hardly any variation in the pattern of church attendance by community size among Roman Catholics in 1995.

Besides denomination, region, and community size, characteristics such as sex, age, and education have also been found to be associated with church-going. Women, for example, have a higher rate of church attendance than men: "In 1990, 40% of Canadian women and 32% of Canadian men attended religious services at least once a month, a narrower gap than five years earlier when 47% of women and 37% of men attended that often" (Bari and Mori, 1991: 23). The gap between women and men became even narrower in 1997 when, according to Gallup Poll, 30 percent of women and 24 percent of men attended church once a week. The greater church orientation of women compared with men has been explained in terms of women's greater involvement in family activities and the intimate relationship between family and church. While church continues to maintain strong ties with family and influences it, its involvement in the occupational sphere of men has weakened considerably (Glock *et al.*, 1967: 43-5). Since both church and women in Western societies are "traditionally preoccupied with conflict-resolving, emotional healing, integrative and expressive functions," women tend to be more church-oriented than men who are "more involved in competitive, differentiating and instrumental activities" (Mol, 1976: 245). Arguments such as these explaining higher religiosity of women than men, based on the traditional notion of sex-based division of labor in society, obviously do not hold true today when almost one-half of the total labor force outside the family is made up of women. The convergence of men and women in labor force participation perhaps explains the narrowing of the gender gap in church participation.

The rate of involvement in church activities tends to increase with advancing age. According to the 1998 Gallup Poll data, the church attendance rates among adults belonging to the age groups 65 and over, 50 to 64, 40 to 49, 30 to 39, and 18 to 29 were 42, 38, 34, 23, and 24 percents respectively. Analysis by Bari and Mori (1991: 22-3) of the general social survey data of Statistics Canada as well as Bibby's (1993: 10) analysis of the Project Canada data on church attendance by different age groups also show rising church-going rate with increasing age. Several reasons have been put forward to explain greater church participation among older people especially in rural areas: a growing concern with life after death; their greater availability to participate in daytime church activities; and their lack of involvement in nonchurch activities (Rogers *et al.*, 1988: 161).

An inverse relationship was observed between education and church attendance in a 1980 Gallup Poll survey: Of those with only an elementary education, 47 percent attended church at least once a week; of those with a secondary education, 34 percent; and of those with a university education, 26 percent. This trend, however, is the opposite of the pattern found in the United States and among Western European countries where a positive association is found between education and church-going. Mol (1976: 246) explained that "in Canada the denomination to which one belongs is much more significant for church attendance patterns than the amount of education one has." The French-speaking Catholics in Canada who are by far the best church-goers, he argued, are overrepresented in the category of people who have only an elementary education but a high church-going rate. Mol's explanation, although valid at the time he offered it, does not hold true any more because, as has been shown above, the church attendance rate among the French-speaking Catholics has shown almost a cataclysmic decline in recent decades. This perhaps explains Bibby's (1993: 10) finding in 1990 that Canadians with a university degree not only had the highest rate of church membership but also had a slightly higher rate of weekly church attendance (27 percent) than those with only a high school diploma or less (25 percent).

Although rural-urban breakdown of the data on the influence of denomination, sex, age, and education on church attendance are not readily available for Canada, it is not highly unlikely that the same trend which characterizes the total Canadian population also holds true for the people in rural Canada. Indeed, the data on rural church attendance in the United States also show greater church attendance among women, older people, and the more highly educated than among men, younger people, and those who are less educated (Bertrand, 1958: 241-2; Rogers *et al.*, 1988: 160-1).

Decline of the Rural Church

Although both urban and rural churches have been affected by the decline in membership and attendance, its effect has been especially severe for rural churches because of their smallness in size. Several factors contribute to the smallness of church congregations in rural areas. Not only the size of rural communities are small but they also tend to have relatively larger number of churches for their population size. "Overchurching" is a traditional rural characteristic especially in those areas where communities are still small in size and sparsely distributed. Denominational heterogeneity and rivalry often give rise to several churches even in a small rural community. In Prince Edward Island, for example, a community of 200 people is sometimes found to have more than one church congregation. A few Protestant families in a predominantly Roman Catholic community may support a church of their own instead of attending a church of the same denomination in a neighboring community. Even in a predominantly Protestant community of small size more than one Protestant congregation are sometimes found to exist side by side.

The traditional problem of overchurching and the phenomenon of rural outmigration have left very few people in rural areas to contribute to the ever increasing cost of maintaining the church. Since the rural church is almost totally dependent on the contribution of its members, the decline in church membership and participation puts the rural church in a precarious financial situation. The diminishing financial support and increasing cost not only have forced the rural church to cut down on its programs but also in many cases close its doors. Rogers and Buckmire (1967: 5-6) describe the process as follows:

> During the first stage the rural minister may leave, and some degree of consolidation may take place in the form of a visiting minister or a student minister. In the second downward step the local population may not even be able to support a visiting minister or a student minister. In the final step the church doors are closed, and the people may have to meet in some other locality or in someone's home. These changes are hastened by a lack of adequate financial support and compounded by the decrease in population.

Most rural churches cannot afford to have the services of a qualified full-time minister. Rural ministers are often low-salaried and inadequately trained; older in age; have less formal education; and have a shorter residency in a given pastorate than urban ministers (Larsen, 1964a: 10-11). Ministers with extensive education and formal training start their career in rural churches only to move on to more lucrative positions in urban churches once they have gained enough experience. On the other hand, older clergymen who may retire to rural areas are often too old and conservative to provide the needed dynamic leadership to the rural churches.

The crisis of the rural church in Canada is exemplified by an account provided by Bishop Fougere (1997) of the problems faced by the 60 Roman Catholic churches in Prince Edward Island. He maintained that "short of a miracle," many would be forced to close their doors because of the rural to urban shifts in population and declining number of priests. The 60 Roman Catholic churches in the province were being served by only 37 priests many of whom were close to retirement. Bishop Fougere estimated that there would be only 18 active priests available in ten years time unless there were new ordinations. Unfortunately, however, there were only four men at various stages in the seminary.

> The reality is, we know, that "short of a miracle" we won't have the people numbers and priest numbers to maintain the physical plants we've had in the past. Some buildings may be closed. A lot of churches were built around farming communities but those demographics are changing. We have no control over who stays in a community. People are going to where there is work and therefore it is likely that certain churches will not have the revenue to maintain buildings.

What does the diocese plan to meet the challenge? It plans to divide the 60 churches into a number of "clusters," each being a grouping of churches and parishes to share a common administration and unite in a pastoral effort. Priests would be assigned to the consolidated clusters of churches rather than to the individual ones. No priest would be required to do more than three Sunday masses on any weekend. When supply priests are not available, those who ordinarily attended a certain church would be asked to attend a neighboring church.

Indeed, "clustering" is one of the six options described by Sim (1990) for the rural churches in Canada to deal with the problems of declining size and attendance. Clustering is defined by Sim (1990: 31) as "an arrangement whereby two or more congregations undertake activities or projects together, whether informally, on a once only basis, or formally with legal or financial glue to ensure continuity." A cluster can be arranged within one denomination — as exemplified by the plan of the Roman Catholic diocese in Prince Edward Island — or among congregations belonging to several denominations. A cluster may thus bring together congregations from one denomination from a wide territory or from several denominations within a smaller circumference. Clustering ensures the availability of resources and services of an ordained pastor to small congregations or of specialized services that single congregations cannot manage by themselves.

The other options, which are not always mutually exclusive, described by Sim (1990) are: community church, part time clergy, small autonomous group, house church, and saddlebag preacher. The "community church" is the local church building used by members of several denominations. It may take several forms: (i) an ecumenical arrangement with members of several denominations participating in the worship together; (ii) one dominant denomination with adherents of other denominations invited for support and joint worship; and (iii) members of several denominations participating in joint worship led by a pastor drawn from each denomination alternately.

A rural congregation is being increasingly served by a "part time clergy" in the sense that two or more congregations share the expenses and benefits of a pastor as in a clustering or multidenominational community church. A "party time clergy" may also be a person who, instead of serving several congregations, has two sources of income, one as a clergy and the other from a second job.

A "small autonomous group" is a congregation whose members "share meaningful religious experience without a church building when too few people and too few dollars are available to keep it going" (Sim, 1990: 29). A small *autonomous* group may also come into existence within a larger congregation made up of individuals "dedicated to worship, study and fellowship together." When this small group of "searchers for spiritual discovery and meaning" becomes more solidly established, they may acquire a more permanent base where it can keep books, materials, and objects of worship giving rise to a "house church." A network of such groups may combine to employ a full or part time professional leader or one or more of their own members receive instruction for "diaconal" type of ministry or, finally, the group may be connected electronically to a great teacher. A small congregation or an autonomous group may also be served by a "saddlebag preacher" who is an itinerant minister making house visitations. The name is taken from the "saddlebag experiment" conducted in Saskatchewan. The saddlebag preacher rarely performs regular preaching except for a special event of festive nature.

A major factor which has contributed to the decline of the rural church is its slowness to adapt to the changing social and economic circumstances in rural areas brought about by the technological revolution in agriculture. The need for the rural church to effectively deal with the changes in rural Canada was emphasized by Mann

(1948) as early as the 1940s. In his book on the rural church in Canada, Mann provided a crisp account of the changing technology and agricultural economy in the countryside and their impact on the rural society. He noted that there was a weakening of moral standards and loss of faith in distinct values of rural living. The old form of rural neighborhoods and bonds among neighbors were in the process of decline. Mann concluded that there was an urgent need for rural churches to deal with these changes. Even the church leaders themselves realize that while the economic, agricultural, social, and educational patterns have changed significantly in several areas over the last several decades, "the only facet of life that has not changed is the church" (Whyte, 1968: 87).

The increasing separation of the sacred and secular domains in rural life, and the inability of the rural church to bridge this gap are also responsible for its declining influence. While the rural life has undergone a revolutionary transformation in its secular sphere, it has remained more or less static in its sacred sphere, and has lost its relevance to the rural people. Whyte (1968: 92) argues that "the lag which the rural church manifests is induced partly by its inherent ideology." The church and the traditional role of its minister represent a "symbol of sacred values" to its members, especially those who are older in age, and any attempt to alter them to adapt to the changing circumstances of rural life are seen "as a threat to the values which they endorse."

Does it mean that the total demise of the rural church is inevitable? Rogers and Buckmire (1967: 7-8) suggest that while there is no one or combination of factors that will provide a solution to the problem, consolidation and re-examination of programs and purposes of the rural church appear to have promises for possible solutions. The solution to the problem of overchurching and diminishing membership, for example, perhaps lies in the consolidation of churches. Congregations belonging to the same denomination could probably be consolidated with least complications. However, experience has shown that consolidation is most difficult in those denominations where congregational autonomy is highly emphasized. On the other hand, consolidation is more easily acceptable to the congregations which are parts of large and strongly hierarchical denominations such as Anglicans, and Roman Catholics. Consolidation of congregations belonging to different denominations is obviously most difficult to obtain. Experience has also shown that for it to be successful, the boundaries of a newly consolidated pastorate must conform to the existing natural community boundaries rather than being arbitrarily arrived at (Larsen, 1964b: 37). Instead of imposing consolidation from above, the idea of a larger parish should come from the people themselves.

In addition to consolidation, the goals and programs of the rural church must relate to the changing needs, attitudes, and beliefs of rural people. The rural church must respond to both religious and secular needs of its members in moral and educational terms. A very important prerequisite for a revitalized rural church is the dynamic leadership of an adequately trained clergyman. Efforts must be made to attract such a clergyman to locate in a rural area. Indeed, as representatives of the rural church, the clergy in the past provided the leadership not only in looking after "the spiritual well-being of their flock" but also "the material needs of those communities" (Jean, 1997: 11). Many of them, especially the clergy of the Catholic Church, were instrumental in creating insurance cooperatives, organizing savings and credit institutions, spreading agronomic knowledge and handling health care, social affairs and public education. The clergy were held in such a high public esteem that large farming families were ready to offer one of their children to the institution and considered it "a tangible sign of respectability and social success" if their offer was accepted. With increasing secularization of the rural society, however, the church has had to withdraw from the varied material and social domains of rural life. The rural depopulation in recent decades, however, has made it impossible for many small rural communities to support "certain professions, artisans and commercial establishments" which have consequently gravitated to larger towns and cities. In view of the fact that the local government is relatively weak in the political tradition of Canada and there is a dearth of local leaders in smaller communities "capable of exercising effective leadership," the responsibility has fallen once again on the shoulders of the rural clergy to "play a major role in the social dynamic" of rural communities (Jean, 1997: 11-2).

There are others (Smith and Zopf, 1970: 358-9), however, who think that the chances of survival of the rural church in the open country is very poor unless conditions change very drastically. According to this view, the prospects of survival of rural churches located in villages and small towns rather than those in the open country are reasonably bright. These churches seem to be better equipped to respond to the changing circumstances of rural life by drawing ideas, talents, and support both from nearby towns and cities as well as from the open country.

THE RURAL SCHOOL AND EDUCATION

Besides providing formal education to children, the rural school traditionally served as a structural focus of the community in which it was located. The boundaries of a school district often delineated the geographic boundaries of a local community or neighborhood. The little white one-room school was not only an educational center but also a place for social and recreational activities.

While the one-room school was very closely integrated with the community, it lacked the resources to meet the demands of a rapidly changing society. One of the significant trends in rural education has been the consolidation of one-room school districts into larger units. Although consolidation has solved many of the problems associated with the one-room school, it has also robbed the small communities of one of their most cherished and important institutions.

The One-Room School

To many of Canada's early settlers formal education was a matter of little concern. Children were socialized in the family to be farmers, fishermen, lumbermen, trappers or hunters. It was not until the last half of the nineteenth century when free and universal education became established in Canada after a stormy beginning. Churches were opposed to secular education supported by the public fund; and many people in rural areas were against paying taxes to support a universal education. The storm, however, subsided by the 1870s and "one-room schools began to sprout like dandelions, dotted around the countryside to serve the next hundred years" (Cochrane, 1981: 11).

The rural schools were literally built by the people of the communities although education departments provided modest grants. Furnishing was often poor and inadequate, and mostly home-made. Some communities in the early years could not even afford the luxury of a properly built blackboard, and used home-made chalks.

The role of a one-room school teacher was most difficult to perform. Since it was not until the 1940s when enforceable standards in teacher contracts was established, the one-room school teacher was under the complete mercy of the school board and the community. The salaries were abysmally low, and female teachers, who earned significantly less than male teachers, were hired by many school boards to save money. The teacher was expected to set a moral example both to the pupils and community residents, and no deviation from community norms was tolerated. Not surprisingly, teacher turnover in rural schools was very high.

Although often seen as "grim men, tightfisted, close minded, bullying teachers for sport" (Cochrane, 1981: 142), the trustees were generally conscientious men doing their best for the school and the community. Elected at the annual meetings of taxpayers to serve a three-year term, they acted as a corporate body and exercised powers vested in them by provincial acts.

The teaching and learning environment in the one-room school was not only affected by the inadequate physical facilities, an inadequately-trained and low-salaried teacher, and hard-fisted and parsimonious trustees, but also by the socioeconomic background of pupils, class-size, teaching methods, and reading materials. Many students performed chores on their farm before they arrived at schools cold and tired walking several miles from the home. The high-absenteeism or dropout rate in rural schools were not a consequence of indifferent attitude toward formal education alone but also of poverty.

The teacher had to spend more time in keeping the noise down, disciplining her pupils, and tending the wood stove than teaching. Since the students were of different ages and at different grade levels, she could spend only a small amount of time with each group of students. This problem, however, was dealt with by adopting the Madras School technique developed for British colonial schools in India (Peabody, 1992: 56). The technique involved using the older and more able pupils to teach the younger and less gifted ones. The students were thus more or less on their own, and only those with a high motivation to learn, worked hard and made satisfactory progress. Only a small percentage of the students went on to high schools in towns. The teacher was often ill-equipped to help students who could not afford to leave home and board in town to attend high school.

The only meaningful contact with the larger world of education was maintained through the school inspector of the provincial department of education who visited the schools in the fall and spring (Cochrane, 1981: 119). A favorable report from the inspector was necessary for continuation of the government grant for partial support of the school.

The school was also a place of social and recreational activities for the community. It was used as a forum for discussing public issues, as a social center, and as a classroom for adult education. The Christmas concert at the school was an occasion of community-wide participation in making costumes for children, building a platform, playing Santa Claus, and giving presents to the teacher and the children. Dollars to buy presents were raised by the whole community in fund-raising parties during the autumn when "men bid on lunches packed by their ladies" or "adults trotted out their songs and recitations for one another's amusement" (Cochrane, 1981: 85).

Problems of Rural Education

Poor school attendance, high dropout rates, and low educational aspiration and attainment have been the traditional problems of rural education in Canada. Many of these problems originated from the inadequately equipped one-room schools which until recently symbolized the low status of education in rural Canada.

School Attendance

Attendance in one-rom schools was appallingly low in the last half of the nineteenth century when they were first started. The record during the first fifty years of the twentieth century was not all that much better. The daily attendance in one-room schools of Prince Edward Island, for example, was often less than 60 percent and hardly over 70 percent until the 1940s (Table 6.4). The attendance figures, however, showed significant increase in the 1960s and 1970s. A rapid transformation took place in the Prince Edward Island school system in the 1960s which not only brought school reorganization in the form of consolidation but also meant increased funding for the remaining one-room schools (Edmonds and Bessai, 1978: 6)

TABLE 6.4

AVERAGE DAILY ATTENDANCE IN ONE-ROOM SCHOOLS IN PRINCE EDWARD ISLAND, 1920-1970

Year	Average Daily Attendance in Percent	Year	Average Daily Attendance in Percent
1920	59.0	1950	78.0
1930	66.3	1960	87.8
1940	68.7	1970	92.3

Source: Edmonds and Bessai (1978: 6).

Student Retention

The student retention rates of one-room schools were even more dismal than their daily attendance rates. Only a very small proportion of rural students continued to grade 12. This was especially true for the Atlantic provinces and Quebec. For example, less than 30 percent of the students who entered grade 2 in 1948 were in grade 12 in 1958 in the Atlantic provinces. In Quebec the corresponding proportion was less than 10 percent. The rate of student retention was also very low in Manitoba, British Columbia and Ontario (Whyte, 1968: 77).

While poor facilities and reading materials, and inadequately trained and overworked teachers were important contributing factors, poor accessibility to high schools, low socioeconomic status of parents, and lack of student interest and motivation were also found to be associated with the low rate of student retention in a study in Saskatchewan (Royal Commission on Agriculture and Rural Life, Report 6, 1956: 235-6). The Saskatchewan study showed that the drop-out rate was higher among children from low income and farm families than those from above average income and nonfarm families. Since the incidence of families with a lower than average income was high among the farming population, and only a few farm parents had a high educational attainment, children from farm families lacked the interest and motivation for formal education. On the other hand, children of parents with an above average income and a high educational attainment tended to be motivated to complete high school. The low retention rate of rural schools were thus perpetuated not only by their inadequate facilities and resources but also by the low income and educational attainment of farm parents.

Educational Aspiration

Very few studies are available on educational aspiration of rural children in Canada. A study conducted in mid-northern Ontario (Blackburn *et al.*, 1975) found that students from farming areas and small rural villages had a significantly lower level of educational aspiration than those from towns and cities. On the other hand, in a study conducted in Prince Edward Island (Dasgupta and Nagarajan, 1978: 26) no statistically significant association was found between educational aspiration and rural-urban residence. There was no significant difference between the urban and rural male students in occupational and educational aspirations. The urban and rural female students, however, were significantly different in both of these characteristics. In the Ontario study, both male and female urban students had a significantly higher level of occupational aspiration than their rural counterparts which contributed to their differences in educational aspiration as well.

Structural characteristics of the rural society and of the family in which rural children are socialized contribute to their low aspiration levels in education and occupation. A narrow educational experience in inadequately equipped rural schools, and a lack of immediately visible nonagricultural occupational roles in a traditional rural society contribute to a low occupational empathy among the rural youth which in turn contributes to their low educational aspiration (Lipset, 1955: 225). Family factors appear to have the most important influence on the educational aspiration of the rural youth. Both Ontario and Prince Edward Island studies show that the levels of education of the father and mother significantly affect the educational aspiration of their children. The Prince Edward Island study also shows that the levels of education of older siblings, parental feelings toward formal education and their expectations, and the level of family income are also significantly associated with the educational aspiration of children. In an earlier study of educational aspiration of high school students in Manitoba, Siemans (1965) found a similar association between educational aspiration, on the one hand, and socioeconomic status of the family and level of education of both father and mother, on the other.

The low level of educational aspiration of the rural youth is thus a function of an educational system weak both in "resource" and "normative" aspects (Whyte, 1968: 68). Although the one-room school and its minimal resources and emphasis on fundamentals provide an education which was perhaps functional in a rural society where subsistent agriculture was the most dominant form of economic activity, it was ill-equipped to raise the levels of educational and occupational aspirations of the rural youth comparable to those of the urban youth. The normative system of the rural traditional society, consistent with its subsistent agricultural economy, placed very little emphasis on formal education. The parental fear that a formal education might take their sons away from agriculture, reinforced the low level of educational and occupational aspirations among rural children. The changed technological and social circumstances of the contemporary rural life, however, have transformed both resource and normative aspects of rural education in Canada.

Consolidation of Rural Schools

The movement to dismantle the one-room school system in favor of consolidation started almost as early as it was established in Canada. To many education officials, the one-room schools were inefficient and wasteful, and failed to offer the kind of education to rural children which was available in urban schools (Edmonds and Bessai, 1978: 131). Some consolidation schemes were carried through by a process of slow growth in the early years of this century but many of them were subsequently dissolved. Rural Canadians viewed consolidation to be expensive and a heavy tax burden, and too far from home for little children to travel by expensive and unreliable transportation (Cochrane, 1981: 158).

One of the early moves for school consolidation in rural Canada came from a private source. Sir William MacDonald, an entrepreneur and philanthropist, financed the building of four model consolidated schools in four provinces to demonstrate their advantages. He even provided the funding to supplement teachers' salaries and cost for transportation of students. Even though these experimental schools demonstrated discernible successes both in attendance and retention rates, and governments took over funding of the consolidated schools past the years of MacDonald's grant, the movement sloughed off by 1920 because of stubborn resistance, and complaint and wrangling over the operation of the schools (Cochrane, 1981: 159).

The technological revolution in agriculture and the resulting prosperity of the 1940s and 1950s contributed to a change in the attitudes of rural people. There was an increased awareness of the gaps between urban and rural people in income, amenities, and services including schooling, and a readiness to attempt to bridge these gaps. Farming came to be viewed as a means to earn a living through a deliberate use of the new and improved technology rather than an occupation to be taken up by every rural young man almost as a religious calling. Increased commercialization of agriculture also made farming an intensely competitive occupation, and only those people who had the requisite skills of modern farming could survive as farmers. Farming ceased to be an "open" occupation to be entered into with minimal skill and formal education. The rural youth with little or no formal education had very few jobs waiting for them outside of agriculture. Rural attitudes toward formal education went through a radical transformation in the changed circumstances of agricultural technology and rural economy.

The stubborn resistance to consolidation started to weaken. The process of consolidation gained some momentum between 1940 and 1945. The rate of consolidation became faster and between the mid-Fifties and mid-Sixties, and by the beginning of the Seventies there were a few one-room schools left in rural Canada (Whyte, 1968: 71). Educators were thus able "to convince a willing public that education was an investment, that the system must be expanded to provide equal opportunities for all, and that modern needs could only be met by expanding the range of school offerings" (Smitheram, 1975: 58). The process of consolidation resulted in a 40 percent increase in enrolment at the elementary and secondary levels. A postconsolidation study in Prince Edward Island showed that retention rates from grade 2 to grade 9 level increased from 69 percent in 1959-60 to 97 percent in 1978-79 after consolidation. The retention rate from grade 2 to grade 12 increased from only 28 percent in 1962-63 to 63 percent in 1978-79 (Smitheram, 1982: 190).

Benefits of Consolidation

Consolidation of schools has brought about a number of desirable changes in rural education. A centrally located consolidated school equipped with libraries, laboratories, and gymnasiums can economically serve a large number of students instead of spreading out these resources thinly among many small schools. Consolidation has made a great variety of learning materials and a wide range of choices of programs available to rural students living even in the most remote communities.

Consolidated schools have qualified teachers who are provided with an improved environment for teaching and their own professional growth. A teacher is not expected to teach a variety of subjects in a class room filled with students at different grade levels. Improved qualification of teachers, better salaries and working conditions, and a more efficient supervision have lowered the rate of turnover of teachers and raised the quality of instruction.

Consolidation has also facilitated the broadening of the curriculum. The curriculum in one-room schools was narrow, limited, and often ill suited to the needs of rural children which perpetuated traditionalism in rural society. In consolidated schools there are ample resources available to develop innovative curricula and teaching methods best suited to the needs of the pupils.

Students in a consolidated school not only benefit from the availability of a variety of program of studies under the direction of a specialist, and improved facilities and learning resources, they also gain from their exposure to a larger student body drawn from many and varied communities of the enlarged school district. This broadens their world of interaction and exposes them to a greater variety of ideas and values. The enlarged size of the consolidated school promotes wholesome competition and stimulus in school work, encourages group and project work, and enriches extra-curricular programs and activities.

Disadvantages of Consolidation

Consolidation has a number of disadvantages as well which have both personal and social implications. School consolidation in rural areas has meant rather long and tiresome bus rides between home and school for a great majority of children. The bus rides are not only costly and wasteful but also have negative emotional and educational effects on younger children. There is some evidence that five and seven-year old children experience a degree of emotional and social maladjustment due to the extended period of absence from home (Lee, 1957). Another study (Dunlop *et al.*, 1957) showed that transported Grade 2 students suffered in educational achievement by comparison with those who did not travel.

The decline of rural schools and their consolidation also have some serious social implications. Small rural communities have lost a very important institution due to the demise of one-room schools. Although the void left by the one-room school has been partly filled by the church and such groups as Women's Institute (Dasgupta, 1984), its disappearance has seriously undermined the small rural community. The consolidation of rural schools has also hastened the process of urbanization and decline of rural communities. Since rural children spend a good part of the year in a school outside of their community of residence, "their community roots become even more fragile. What is much more serious is that because the consolidated school setting is built around urban values, rural children are being taught to become urbanites" (Smitheram, 1975: 69-70). The reading materials are filled with urban models of social and economic living, and the curricula of rural schools are little different from those of metropolitan schools. Randhawa (1992: 110) goes as far as to claim that the closure of small rural schools through consolidation has contributed to the rapid decay of many rural communities in the prairies and elsewhere turning them into "ghost towns." Peabody (1992: 137) is less dramatic in his assertion regarding the negative effect of consolidation on rural communities. While he concedes that consolidation has enabled rural schools to provide a standard of education to their students equivalent to that provided by urban schools, he wonders about its effect on the long term viability of the rural community. If the children have to go to a different community each day to attend school, they will come to believe education and other social and economic opportunities are not to be found in their own communities. When the children of a community come to believe that their "community holds no future for them, then the community itself has little hope to avoid decline."

A major source of local distrust of consolidated schools is the decline of local control on schools, and the feelings among parents that their involvement in and influence on school-related activities have declined. Board members of consolidated districts are not as easily accessible to parents as those of small school districts, and the school system has become too large and bureaucratically too complex for rural parents to have any direct control on them.

Debate Over School Consolidation

Although by the 1990s there was hardly any one-room schools left in Canada, the debate over school consolidation has not ceased. The debate is essentially between "social localists" and "social expansionists" (Brown, 1968: 4). The concern of social localists goes beyond the question of school consolidation. Their struggle is against the gradual erosion of the local autonomy not only in the sphere of education but also of economy, religion, and recreation. They hold the view that the process of reorganization and centralization of services is disintegrating the local community.

The concern of social expansionists, on the other hand, is the rural-urban lag in educational as well as in other spheres of society. They view the process of homogenization as a means for equalizing opportunities for both rural and urban populations to improve their standard of living. Studies conducted in the 1950s provided compelling arguments for the reorganization of schools in rural areas. A comparative study of educational attainment in rural British Columbia (Andrews, 1954) showed that although both groups of students did equally well in reading, writing, and arithmetic, the curricula of the reorganized schools were broader and more enriched. In a series of studies comparing scholarly achievement of elementary and junior high school students from urban, town, graded rural, and ungraded rural schools in Alberta (Dunlop *et al.*, 1955), the pupils of the ungraded rural schools were found to have the lowest achievement tests scores. The social expansionist view was emphasized by Card, an educational sociologist, in his brief to Alberta Royal Commission in 1958 (Brown, 1968: 7) that "rural youth migrating to the city from educationally undernourishing rural schools will accept, and in fact must accept, a lower standard of living than that which they left."

Social localists concede that rural children need a better education but are not convinced by the argument that a good education is only possible in large consolidated schools and that the latter is the least expensive way of providing a better education ro rural children. They argue in favor of both "the maintenance of the community school and the improvements proffered by consolidation" (Smitheram, 1982: 197).

There are indeed some empirical support to this view of social localists. In a 1973 study of 29 one-room and 27 two-room schools in Prince Edward Island, Edmond and Bessai (1978) found that there was no significant difference between the better equipped and better staffed small schools and large consolidated schools in the educational attainment of their students. The small schools were superior to the large schools in learning environment and interpersonal relationships as evident from the opinions of teachers and students surveyed. Most teachers, however, rated the large schools to be superior in terms of reference materials, availability of audio-visual aids, and extracurricular activities. The difference between the small and consolidated schools was most significant in the social climate. The students of the small schools consistently rated their schools higher in satisfaction and cohesiveness, and lower in friction, competition, and difficulty in work than those of the consolidated schools.

The argument that reorganization of small schools into large unit involves substantial cost savings is clearly supported by scale studies for secondary school facilities but findings on elementary schools in Ontario show little evidence of economies of scale (Dawson and Dancey, 1976). A study of the scale economies for rural elementary schools in Prince Edward Island (Keane, 1978) showed that although there were economies of scale in the operation of consolidated elementary schools, the economies were not very substantial, and could be offset beyond a certain size by other factors such as rising transportation costs. The author argued that the findings of his study suggested that while consolidation at the elementary level resulted in insignificant cost saving, it indeed might have raised the social cost which is more difficult to measure. His study, for example, indicated that "areas with small schools have fared much better in terms of their ability to retain population than similar areas without schools."

It will be fair to conclude that the findings of the studies on the effects of school consolidation are far from conclusive. The debate over school consolidation is not only a debate between two sides with opposing educational philosophies but also with conflicting views regarding the cost the society should be asked to pay for an improved quality of education in rural Canada. While social expansionists find the small rural community structure expendable for the equalization of educational opportunity, social localists insist on both improving the quality of rural education and maintaining the identity of the rural community and its values.

A graduated or staggered consolidation has been suggested by some as a compromise between the two conflicting demands (Smith and Zopf, 1970: 333-4). The small local community is the natural unit best suited to serve as an attendance area for a school including the first three to six grades. By the time the child is old enough to enter grade 7, the child is physically more sturdy to withstand the stress of a long bus ride, and mentally more mature to deal with a wider and somewhat impersonal social world and take advantage of the varied resources offered by the consolidated school. The experience in a consolidated school at higher grade levels prepares the child furthermore to enter community colleges located in larger towns, and later to go to universities located in cities. This gradual consolidation capitalizes on the natural social units in rural areas rather than viewing them as expendable entities, and enables small rural communities and larger villages and towns work together as focal points for more improved educational activities.

THE RURAL LOCAL GOVERNMENT

The people in rural Canada are affected by two general types of governments: (1) federal and provincial, and (2) local. These two broad categories of government, however, are interrelated. The federal and provincial governments through their administrative, legislative, and judiciary machinery help relate local governmental units to each other, and to the overall national and provincial structures. The local governments on their part influence federal and provincial units through the political processes at local levels.

The involvement of rural people in the affairs of the structurally remote and complex federal and provincial governments is indirect and superficial. Their participation in the processes of local governments, however, is more direct and intense because they actually come in contact with political institutions and agencies at this level and see them function. The local government deals with matters of immediate local relevance and concern which can receive only scant attention at the provincial and federal levels.

Traditionally, rural local governments enjoyed a higher degree of autonomy than their urban counterparts primarily because of their relative isolation. In recent decades, however, the autonomy has diminished significantly as rural communities have become more easily accessible to urban centers. Rural local governments were also more informal and personal both in their form and function because of their greater reliance on local traditions and customs rather than on formal laws and ordinances. These informal qualities provided them with a degree of stability but made them less responsive to sociocultural changes. Rural local governments also suffered from the absence of an efficient and specialized political leadership either because of lack of people with such qualities and experiences in rural areas or of any material incentive for such people to be drawn into positions of leadership in local governments.

Development of Local Government

The history of the local government in Canada is little over a hundred years old. It was not until the middle of the last century when province-wide systems of local governments were established. Inspite of its relative youth, the history of local government in Canada is rather complex (Rowat, 1975: 1-7; Crawford, 1958: 10-47).

The old system of local government was "Government by Session." It was, however, not a *local* self-government but an administrative *extension* of the central government into local communities. The government was made up of two groups of officials: Justices of the Peace and a Grand Jury to assist them. They were local landowners, merchants, and lawyers appointed by the central government as its local representatives to see that things were done according to law. They also made local regulations, licensed tavern-keepers, controlled roads, bridges, ferries, and fisheries, supervised the jails and the police, and regulated marketing and standards of agricultural and forest products.

The Government by Session form of local government lasted for almost a century in Canada. It was a local government imposed from the above over which local taxpayers had very little influence. Although the Government by Session was the predominant form, rural self-government which grew up in England through the centuries also started to make its appearance in Canada in the form of the Township in Ontario and the Parish in Quebec. The "township" form of local government was first established in New England and later entered Canada with New Englanders who settled in the Annapolis Valley of Nova Scotia in 1760 (Rowat, 1975: 6). The New Englanders, however, failed to set up township as an independent, self-governing unit in Nova Scotia but succeeded later in Ontario and Quebec.

In the early 1840s the Courts of Session were replaced by elected District Councils in rural areas of Ontario and Quebec. Since District Councils proved to be too large, they were replaced by parish and township governments in 1845, but these in turn were replaced by County governments in 1847. A two-tier system of local self-government was finally introduced in Ontario in 1849 and in Quebec in 1955. In addition to the County government, small self-government units were established for cities, towns, and villages within boundaries of the counties. In Quebec the units under the county were parishes, townships, towns, and villages. Except for the recent creation of metropolitan governments, the two-tier system of local governments continues to exist in Ontario and Quebec even today.

Local self-government did not come into existence in the Maritimes until 1877 when in New Brunswick and Nova Scotia rural areas were organized into incorporated counties. Local governments at the county level were never established in Prince Edward Island.

Since the principle of local self-government became well-established in Canada by the time Manitoba was created in 1870 and Saskatchewan and Alberta in 1905, the establishment of local self-governments in the three Prairie provinces involved no resistance After an initial attempt to implement it, Manitoba abandoned the county model of local government because of the sparseness of its rural population, and established self-governing units in cities, towns, villages, and rural municipalities in 1902 which continue to exist today. Local self-governments were established in Saskatchewan and Alberta even before they became Provinces. Besides creating self-governing units for towns, rural areas were organized into municipalities but called "counties" although they were not the same as the two-tier counties of Ontario and Quebec. As in the Prairie provinces, the central Canadian county system could not be established in British Columbia because of the sparseness of its population. Self-ruling villages and towns were established much later although in 1872 a general act made provision for the voluntary creation of local units.

The form of local self-government has remained fundamentally the same even after more than a century since it was first created. The durability of its present form either reflects the foresightedness of its creators or very little interest among people to experiment with the structure of local government. According to Rowat (1975: 13), the time has come for a major reorganization of local governments to effectively relate them to the vast social and economic changes in Canada.

Contemporary Local Governments

There are two broad kinds of local governments in Canada: (1) municipal governments, and (2) agencies, boards, and commissions. The municipal government exists at the level of a county, city, town, village, parish, township, or community.

The local government in the form of an agency, board, or commission (ABC) functions within the same local area of a municipality or that of a small group of municipalities with its own separate board and enjoys a degree of autonomy to provide a specific service within the area of its jurisdiction (Siegel, 1994: 7). School boards, park commissions, housing authority, business development districts, and downtown development corporations are some of the examples of the ABC type of local governments. The ABC governments, however, are found mostly in urban areas although the agencies like school boards and health commissions may include rural areas within their jurisdiction as well.

The local government in rural areas is represented primarily by the municipal government. Municipalities are created by provincial governments who also determine their powers and responsibilities. Municipal governments show a wide variation in size, form, and function from province to province because they have more to do with precedent and tradition in the provinces rather than with rational considerations (Diamant, 1997: 69). There are three types of municipal governments in Canada: (1) "unitary municipalities," which are local units at the levels of cities, towns, villages, townships, parishes and communities; (2) "regional municipalities" with larger areas under their jurisdiction including local municipalities within their boundaries; and (3) "quasi-municipalities," which are nonincorporated units with limited power and functions. Table 6.5 shows different types and numbers of unitary and regional municipalities in different provinces. It should be pointed out, however, that the numbers presented in the table are close approximations rather than exact. The myriads of types into which the municipal units are divided in each province makes it difficult to summarize them into a few categories with precise number in each category.

Unitary Municipalities

Unitary municipalities in different provinces are classified into four types: city, town, village, and rural. The first three are urban municipalities although the population of an incorporated "village" could be as small as 100. In Newfoundland and Prince Edward Island the smallest municipal units are called "communities" rather than villages. The category "village" in Table 6.5 also includes "northern villages" in Saskatchewan and "summer villages" in Alberta which often have fewer than 100 residents.

TABLE 6.5

TYPES OF MUNICIPALITIES BY PROVINCES IN CANADA, 2000

Municipalities	Nfld.	PEI	NS	NB	Que.	Ont.
Unitary	269	76	52	109	1367	557
City	3	2	--*	6	273	48
Town	136	7	30	26	--**	121
Village	130**	67***	22	77	135	45
Rural	--	--	--	--	959	343
Regional	--	--	24	--	99	49
Metropolitan	--	--	3	--	3	3
County and District	--	--	21	--	96	46

Municipalities	Man.	Sask.	Alta.	BC	Canada
Unitary	198	796	368	145	3937
City	4	12	13	44	405
Town	41	145	109	15	630
Village	37	342	166	40	1061
Rural	116	297	80	46	1841
Regional	2	--	--	27	201
Metropolitan	--	--	--	--	9
County and District	2	--	--	27	192

* The three cities of Nova Scotia were merged with neighboring towns and rural municipalities creating new regional municipalities in the mid-1990s: Halifax and Dartmouth became parts of Halifax regional municipality and Sydney became a part of Cape Breton regional municipality.

** The major cities are parts of regional metropolitan municipalities. Towns are grouped together with smaller cities because they are identical in their forms and powers.

*** In Newfoundland and Prince Edward Island smaller incorporated units are designated as "community" rather than "village."

Source: Based on information presented in *Canadian Almanac and Directory 2000* (1999: 4.1-4.61).

There is little uniformity from province to province in the minimum population required for each type of municipalities. The required minima of city municipalities range from a few hundred to 15,000 in different provinces. The population range is between 500 and 1,500 for towns, and 100 and 750 for villages although there is no minimum population requirement for villages in New Brunswick and communities in Prince Edward Island. It is thus not unusual to find villages that are larger than towns, and towns that are larger than cities in different provinces (Rowat, 1975: 17).

Not all provinces have all different types of unitary municipalities. Newfoundland and Prince Edward Island, as noted above, have communities rather than villages. Rural municipalities do not exist in Newfoundland, Prince Edward Island, and New Brunswick. After the amalgamation of the three cities in Nova Scotia — Halifax, Dartmouth, and Sydney — in the 1990s with regional municipalities, the province does not have a separate city municipality. Since there is no distinction between a city and a town in Quebec in terms of jurisdiction power, the smaller cities and towns are grouped together under the category "city" (Table 6.5) The rapid growth of population surrounding the large cities have also blurred the distinction between rural and urban municipalities. Many formerly rural municipalities have grown into densely populated units. The lack of uniformity in the definitions of local municipal units thus makes interprovincial comparison difficult.

Regional Municipalities

Regional municipalities have much larger territorial jurisdictions than unitary municipalities. They are found in five provinces. The regional municipalities in Nova Scotia and Manitoba do not constitute second tier municipal governments as they do in Quebec, Ontario, and British Columbia.

Of the 21 regional rural municipalities in Nova Scotia, eight are at the county level and 13 are at the district level. The district level rural municipalities were created by dividing the larger counties into municipal districts. Besides the 21 regional rural municipalities, three additional regional municipalities were created in Nova Scotia in the 1990s amalgamating cities, neighboring towns, and rural municipalities. The unitary municipal units such as a town or a village located within the territory of some of these regional municipalities maintain their separate identity and autonomy and are not considered subordinate to the larger units although they are free to cooperate with each other to provide services to the residents. The two regional districts in Manitoba and the unitary municipalities located within their boundaries also hold similar relationship.

The regional municipalities of Quebec, Ontario, and British Columbia represent second tier municipal units. They contain unitary municipalities within their boundaries as subordinate units. While the first two provinces have both metropolitan and county municipalities, British Columbia has only rural districts as regional municipalities. The regional municipal councils in these provinces are often made up of representatives of the subordinate rural and urban unitary municipalities located within the territory of their jurisdiction.

The single-tier municipalities, as opposed to two-tier, has been the most common and enduring type. They are more popular because residents find the single-tier form of municipal government most accessible and responsive to their needs: they "feel free to, and comfortable about phoning their councillor to raise an issue, request a service or complain about a decision of council" (Diamant, 1997: 67). The two-tier system is often too large and complex for residents to know clearly "who is responsible and accountable for problems that arise and decisions that are made." Although the two-tier system provides a framework for smaller municipalities to formally cooperate to deliver services which they cannot provide on their own because of smaller population base and limited resources, they are not common predominantly for two reasons: (1) the small, single-tier unitary municipalities are popular because even though they may not have the resources to "deliver a comprehensive array of services but those that are delivered are responsive to local needs"; and (2) the usefulness of the two-tier system in smaller provinces is questionable because the latter lack the population base and density to take full advantage of the system's potential (Diamant, 1997: 67-8).

Quasi-Municipalities

Quasi-municipalities have limited power and functions. They are often small improvement districts created to provide certain municipal services to their residents. According to one estimate (Statistics Canada,1989: 19-28), there were over two hundred quasi-municipalities in Canada in 1988. They were found in largest number in Newfoundland (143) while Ontario (8), Manitoba (17), Saskatchewan (14) and Alberta (19) had a small number of quasi-municipalities.

Rural Municipalities

There is a great diversity in the organization of rural municipalities in different provinces (Table 6.6). Newfoundland has no rural municipal organization excepting local government communities in smaller settlements with limited power and functions. The rural areas of Prince Edward Island are governed directly by the province except the local areas designated as communities with councils of their own to provide certain municipal services. Rural municipalities are absent in New Brunswick as well although unincorporated local service districts provide services of a municipal nature. Nova Scotia is the only province in Atlantic Canada which has municipalities in rural areas. The rural municipalities in Nova Scotia are large regional municipalities at the county and district levels. While the smaller counties have municipalities of their own, each of the larger counties has been divided into two districts constituting separate municipalities. Three of these rural municipalities, however, have been amalgamated with neighboring cities, towns, and villages in the 1990s to create three regional municipalities two of which include at least one city. Both Manitoba and Saskatchewan also have unitary rural municipalities as also Alberta where they are designated as municipal districts.

Quebec, Ontario, and British Columbia, the three provinces with the largest populations, have two-tier municipal governments. Metropolitan areas in Quebec and Ontario that have been consolidated by amalgamation or annexation have also evolved two-tier systems to deliver regional services. Rural local government units in Quebec exist in the forms of parish, township (or canton) and municipality. Townships and municipalities represent the basic units of rural municipal governance in Ontario. The basic rural municipal unit in British Columbia is called the district municipality.

TABLE 6.6

TYPES OF RURAL MUNICIPALITIES BY PROVINCES IN CANADA

Province	Rural Municipalities	
	Basic Unit	Second Tier
Newfoundland	None	
Prince Edward Island	None	
Nova Scotia	County and District	
New Brunswick	None	
Quebec	Township, Parish and Municipality	County
Ontario	Township and Municipality	County, District, and Regional Municipality
Manitoba	Municipality	
Saskatchewan	Municipality	
Alberta	Municipal District	
British Columbia	District Municipality	Regional District

For the rural areas of Quebec, the regional county municipalities constitute the second or upper tier of rural municipalities. Townships, parishes, and municipalities located within a county are subordinate units of the regional county municipality. The council of a regional county municipality includes representatives from all local or basic units subordinate to it. The unincorporated areas within a county are directly administered by the council of the regional county municipality. County, district, and regional municipalities constitute the second tier rural municipalities in Ontario. A county or regional municipality in Ontario is a federation of local units situated within its boundaries although cities and separated towns do not come within the administrative jurisdiction of its council. District municipalities in British Columbia are subordinate units of a more comprehensive form of municipal government called the regional district. The regional district as a federation of local units has administrative authority not only over the district municipalities but also over the unincorporated areas located within its boundaries.

Types of Rural Municipalities

Various rural municipalities in Canada can thus be categorized into following types: township and parish, municipality, district, and county (Table 6.6). Although not always rural, incorporated villages and towns have important influences on rural areas surrounding them.

Township and Parish. Townships were in existence in Ontario even before the province was established in 1791 (Ross, 1962: 28). Quebec has both townships and parishes as local municipal units in rural areas. A township in Quebec is also called a canton.

Municipality. In Manitoba and Saskatchewan, municipalities are the only form of rural local government. Besides township and parishes, Quebec also has municipalities as the basic unit of governance in rural areas. Municipal units at the level of counties and districts in Nova Scotia are called rural municipalities although they include much larger areas than local municipalities in other provinces.

District. Districts are basic units of rural government in Alberta and British Columbia called municipal district and district municipality respectively. The district municipalities in British Columbia are parts of larger regional districts in a two-tier system. Districts represent upper or second tier municipal units in Ontario as well. In Nova Scotia, larger counties have been divided into districts which have municipalities of their own.

County. County governments are not widespread in Canada. They constitute the basic unit of rural municipal government only in Nova Scotia where, as noted above, there are rural municipalities at the county level. In Quebec and Ontario they represent second tier municipal governments and include lesser units of local government within their jurisdiction.

Incorporated Community, Village, and Town. Incorporated communities, towns, and villages are important units of rural social organization even though they are generally considered as urban municipalities. Indeed, incorporated communities and villages exist as nuclear entities in the countryside providing the rural residents with basic goods and services. Towns serve as trade centers for the rural areas which surround them. All provinces with the exception of Newfoundland and Prince Edward Island have incorporated villages. They occur in largest numbers in Saskatchewan, Alberta, and Quebec (Table 6.5). Newfoundland and Prince Edward Island have incorporated communities rather than villages. Towns are found in all provinces although in Quebec they are lumped together with smaller cities as municipal units.

Municipal Councils

The structure of rural local governments across Canada is essentially similar because even though they are guided by varying provincial legislations, the same philosophy and tradition have formed the basis of legislation in each province. All municipalities in Canada are, for example, self-governing and controlled by municipal councils elected by citizens. The majority of the municipalities in Canada have the Mayor-Council form of government (Rowat, 1975: 23). The Mayor is elected directly by the entire municipality, and as the chairman and chief executive officer of the council he or she wields special powers. The chief executive is usually called Mayor in urban units and Reeve in rural units. In the rural units of Quebec he or she is called Mayor rather than Reeve, while in those of Nova Scotia, and in the second tier counties of Ontario and Quebec, the term Warden is used (Rowat, 1975: 23). County Wardens, however, are not elected directly by citizens but by county councillors from among themselves.

Councillors are elected by districts or wards in larger municipalities, and by the entire unit in smaller municipalities. The size of municipal councils varies according to the size of municipalities: while townships and parishes generally have from four to six councillors, the county based rural municipalities in Nova Scotia have as many as 20 or more. In most rural municipal units the term of office for councillors is from one to four years. In many municipalities the whole council is elected anew while in others, such as those in the Prairie provinces, councillors are elected with overlapping terms.

The Mayor, Reeve or Warden is the chairman of the council and acts as its leader by guiding council's discussion to arrive at decisions and seeing that its decisions are implemented. The council itself strikes several standing committees with its own members to deal with its specific functions. In large rural municipalities, the entire council meets only a few times in a year while its standing committees made up of fewer people meet much more often to deal with their businesses.

Functions of Rural Municipalities

According to Rowat (1975: 52), the functions of local governments can be described primarily as "domestic." The domestic functions involve regulating conducts of individuals for the welfare of the community at large. The municipal Acts contain long lists of specific kinds of regulation ranging from "restraining the 'going at large' of domestic fowls" to licensing businesses. Maintenance of roads and bridges has traditionally been the responsibilities of rural governments. Education has also been a local responsibility even though locally elected school boards have generally been a separate local government entity from municipal governments. Provincial governments have increasingly taken over the support of rural schools directly as they have been consolidated and the number of school districts has dramatically diminished in size. Declining rural population and increasing cost of municipal services have made rural local governments more dependent on provincial governments than ever before.

Financing Local Governments

The major problem which plagues both rural and urban local governments is financial. While local governments have to provide a much greater number of services to their taxpayers than before, and the costs of providing such services have increased astronomically, their income has grown at a much slower rate. Because of the diminishing population size and the sparsely distributed population, rural local government have a special financial difficulty. It is much more expensive to provide municipal services in a widely dispersed area of a rural municipality than in an urban municipality. Rural municipalities thus find themselves in a "double bind." While they need to overcome the cost problems of small scale, their overall costs are rising faster than their tax revenues (Rainey and Rainey, 1978: 127).

Local Government Revenues

The problem of local government finance in Canada is twofold: it generates only a half of its total revenues from its sources and depends on the senior governments for the rest; and the real property tax is the only internal source of revenue of any significance.

More than two thirds of the revenues which originate from internal sources of local governments come from the real property tax. This is especially true for rural municipalities because they have little or no income of any significance from other sources. Rural municipalities unlike their urban counterparts collect almost nothing from two lucrative sources of revenues, amusement and sales taxes, and public utilities.

The continued near-total dependence on the real property tax has created most of the recent financial woes of the municipal governments in Canada. The real property tax was a convenient source of municipal revenue in a predominantly agrarian society when property was owned by most people and the limited range of services provided by the municipalities were considered services to property. Property tax is no longer adequate to produce enough revenues for the local governments. An increasing proportion of people rent rather than own their properties, and an increasing number of municipal services are considered services to people rather than to property. The real property tax is regressive in nature because it is based on the assessed value of a person's property rather than his or her income, and thus does not reflect the person's ability to pay (Tindal and Tindal, 1979: 36-8). Finally, the real property tax is slowest in growth rate compared to other sources of revenue, such as income tax, export tax, natural resources revenue and so forth. These sources of revenue are either shared by or reserved for the federal and or provincial governments, and local governments do not have any access to them. Local governments in Canada are thus left with the real property tax which is increasingly unable to meet their growing revenue needs. In 1930, for example, four-fifths of the total municipal revenues was generated by the real property tax but by the 1980s the proportion came down to less than one-third (Canadian Federation of Mayors and Municipalities, 1976: 25).

"*Puppets on a Shoestring.*" The short-fall of municipal revenues is being met by the rapidly growing intergovernmental transfers. These intergovernmental transfers have two characteristics: They are mostly provincial grants, and are meant for specific purposes. These specific purpose grants are based on the priorities of the senior level governments rather than of local governments and thus have turned the latter into puppets with little autonomy. The argument provided in favor of specific purpose grants is that many traditional local government responsibilities such as education have now a wider rather than local significance which need to be exercised at least at a minimum standard across a province. The specific purpose grants seem to ensure the maintenance of such a universal standard (Tindal and Tindal, 1979: 38). The growth of the number of conditional grants to ensure the provision of services at a minimum standard, however, has increased the involvement of provincial governments in local government affairs. Especially in such areas of responsibility as education, roads, health and welfare, the trend is shared involvement of local governments with those at provincial and federal levels. The increasing intertwined operation of local governments whether with those of provincial and/or federal governments has resulted in the extinction of local autonomy in all but a few areas of municipal responsibility.

Problems of Local Government

Besides the problem of financial difficulties, local governments of Canada suffer from other inadequacies as well. These inadequacies both originate from and contribute to the financial problems.

Smallness of Size

The smallness in size, especially in rural areas, has been a major contributing factor to the financial weakness of the local governments. There were over four thousand unitary and regional municipalities in Canada in the year 2000 almost one-half of which were rural municipalities, counties, and districts, and another one-fourth were incorporated communities or villages. In other words, three-fourths of all municipalities in Canada were either rural or incorporated communities or villages, most of which were relatively small in size with a sparsely distributed population. While the postwar rural-urban migration increased the size of many of these municipalities adjacent to urban centers, those which are located further away from urban centers have suffered from the loss of population. The smallness of the population size and the limited assessment base of most of these municipalities not only inhibited the effective provision of services but also made it difficult for them to hire qualified personnel and expensive equipment increasingly required.

Decline of Authority

The financial inadequacy brought about by the smallness of size and declining assessment base, and increasing service demands have caused a significant erosion of the authority of local governments. They lost considerable authority and jurisdiction as separate boards and commissions increasingly deal with specific services which were traditionally under their jurisdiction. Most of these boards and commissions are autonomous entities often created by the provincial rather than the local governments themselves. As noted above, increasing dependence on the funds provided by the provincial governments also contributed to the erosion of the local government autonomy. Although the provincial money made the maintenance of the needed municipal services at a minimum standard at the local level possible, it also substituted provincial priorities for those of the local municipal councils and their citizens (Tindal and Tindal, 1979: 47-8).

Increasing Politicization

According to the Tindals (1979: 46), the increasing politicization of local government is "perhaps the most significant development arising from the postwar pressures of change." The local government in Canada has been traditionally viewed as an administrative rather than a political entity even though politics has always played some part in its operation and decision-making processes. The municipal council, for example, was expected to depend solely on technical and financial criteria in its policy formation and implementation keeping political considerations entirely out of the picture. There was little or no public discussion on alternative courses of municipal action, and "this was considered as natural and desirable." The situation, however, has changed drastically in recent decades with the increasing awareness in the people's mind of the social and environmental implication of developmental activities. There is a growing rise of awareness in rural areas, for example, about the desirability and validity of the growth of new service demands. There is an increasing philosophic controversy regarding the quality of life, and the style and substance of government between farm and nonfarm people, old and new settlers, permanent and seasonal inhabitants in rural areas. Citizens' groups with varied ideas of alternative options are increasingly confronting the municipal councils to influence their decisions. Municipal councils often find themselves ill-prepared to deal with this increased political consciousness of their citizens. The growing politicization of the local municipal activities is further evident in the fact that the candidates seeking office in municipal councils are increasingly coming from political parties or local citizens' groups.

Inadequate Internal Organization

The Tindals (1979: 47) also point out that the internal organization and operations of municipal government are unsatisfactory in "leadership, planning and priority setting, coordination and accountability, and responsibility." Municipal councils suffer from a lack of central authority to take responsibility for progressive action and legislation. The heads of most municipal councils have limited executive power, and have to rely heavily on their personality and persuasive skills to enlist their councils' cooperation. Councils themselves are made up of a motley grouping of people with disparate concerns and interests without any sense of cohesion or collective will. Decisions on planning and priorities in council meetings, as a result, are based not on consensus but on "trade offs." The shortness of the term of office of councillors also discourages the councils to undertake long-term planning. The lack of cohesion of the councils stands in way of effective coordination of activities as well. The standing committees of the councils responsible to oversee the specific municipal programs tend to be overly preoccupied with their particular tasks without making any effort to coordinate each other's activities. The absence of centralized authority, cohesion, and coordination contributes to a serious lack of accountability within the municipal councils. It is often difficult to identify a person or a group of persons who is responsible to take initiatives to deal with the municipal problems or scrutinize and evaluate the initiatives taken by a municipal council. In short, "the chances of buck-passing in this system are endless."

Local Government Reform

The problems of the local government in Canada can thus be listed as follows: (1) insufficient revenue sources contributing to the increasing dependence on the provincial governments for financial support and consequent loss of local autonomy;

(2) smallness of size and the declining assessment base, particularly in rural areas, causing increasing financial insolvency; (3) decline of authority and power due to jurisdictional fragmentation caused by the growing creation of special purpose boards and commissions, and increasing dependence on the senior governments; (4) rising politicization of municipal affairs brought in by growing consciousness among the residents of social and environmental implications of municipal activities; and (5) inadequacies of an internal organization inherited from a century old reform era to deal with problems of the present day local governments.

There have been several attempts to reform the local government in Canada (Tindal and Tindal, 1979: 49-70). They have primarily emphasized administrative and financial efficiency to deal with the problem of smallness and fragmentation of local units, increasing financial burden, and inefficient delivery of the municipal services. But the reforms have been deficient in resolving the problem of fragmentation of responsibilities among municipalities, special-purpose boards, and provincial governments, and in strengthening the revenue sources of local governments to enable them to maintain some degree of autonomy in authority and power. Finally, there has been very little reform effort to improve the internal organization of the local government from the point of view of effective leadership, coordination of municipal activities, and effective representation of local citizens in municipal affairs. The scope of the local government reform has been narrow in the sense that while it has attempted to provide the major municipal services universally at a minimum standard, it has largely ignored the problem of the declining autonomy of local governments.

SUMMARY

The influence of the church and religion in rural Canada has declined with the technological development in agriculture, and increasing secularization of rural life. This is evident in the decrease in the rate of church attendance, and a growing perception among rural people of the declining influence of religion in their lives. The shrinking population base of the rural society, continued overchurching, increasing cost of maintenance, and an ineffective leadership have contributed to a decline of the number of churches in rural areas. The membership in the churches located in large villages and small towns has grown at the expense of those in the open country.

The one-room school which dotted the countryside in Canada, and symbolized rural education, is now a thing of the past. The consolidation of rural schools has meant an educational system in rural Canada comparable to what has been in existence in urban areas. It has meant a higher student retention rate, a broader curriculum, better teaching, and improved instructional facilities. The higher educational attainment of rural students is not only a function of improved material resources but also of a more positive attitude of rural people toward formal education, a result of the changed technological and economic circumstances in rural Canada. Whether the school consolidation has been of beneficial consequence to the rural society, however, is still a matter of controversy.

The local government in rural areas has lost its autonomy because of the same forces which transformed the roles of the rural church and school. The shrinking tax base and increased need of services has made the rural local government financially dependent on the federal and provincial governments. The dependence on the senior governments for financial resources have increasingly made the local government act like an extension of the provincial government in rural areas rather than as local self-government.

CHAPTER SEVEN

Rural Canada in Transformation: Summary and Conclusion

The purpose of this book has been to examine the ways in which the process of social change has transformed the rural life in Canada. A comparison of the traditional and contemporary rural life in Canada (Table 7.1) shows that the rural society has moved increasingly toward the urban end of the rural-urban continuum in demographic, economic, and social characteristics.

TABLE 7.1

SUMMARY OF CHARACTERISTICS OF TRADITIONAL AND CONTEMPORARY RURAL CANADA

Characteristics	Traditional	Contemporary
Demographic		
Proportion of people living in rural areas	High	Low
Proportion of rural people living in non-farm areas	Low	High
Proportion of "farm" people in rural areas	High	Low
Rate of outmigration of rural people to urban areas	Low	High
Dependency ratio	High	Low
Sex ratio	High	Low

TABLE 7.1 (cont'd)

Characteristics	Traditional	Contemporary
Proportion of married people	Low	High
Proportion of people separated or divorced	Low	High
Proportion of people with a formal education	Low	High
Ethnic composition	Highly homogeneous	Some heterogeneity
Proportion of people in nonagricultural occupations	Low	High
Proportion of people in farming	High	Low
Agricultural		
Productive efficiency of farmers	Low	High
Use of labor in agriculture	High	Low
Use of family labor in agriculture	High	Low
Use of hired labor in agriculture	Low	High
Use of machinery in agriculture	Low	High
Use of fertilizers and chemicals	Low	High
Size of farm	Small	Large
Number of farms	Large	Small
Proportion of agricultural product sold in market	Small	Large
Crops grown on the farm	A large variety mainly for domestic consumption	A few mainly to sell in market
Method of marketing the agricultural product	Sold by the farmer directly to the consumer	Various middlemen sell agricultural products in a highly enlarged market
Contract farming	Non-existent	Increasingly important
Willingness on the part of farmers to use modern agricultural innovations	Low	High
Role of agrifood	Almost non-existent	Increasingly important
Role of the "agricultural ladder" as a process to become a farm operator	Very important	Almost non-existent
Proportion of farm owners among agriculturists	High	Low
Proportion of part owners among agriculturists	Low	High

TABLE 7.1 (cont'd)

Characteristics	Traditional	Contemporary
Proportion of tenants among agriculturists	High	Low
Proportion of incorporated farms	Almost non-existent	High
Proportion of small family farms	High	Low
Importance of cooperatives in agriculture and rural life	Low	High
Role of farmers' organizations	Almost non-existent	Increasingly important
Social		
Importance of the small rural community to its residents	High for both economic and social purposes	Low for economic purposes but high for social and emotional needs
Basic unit of rural community organization	Town and country community	Rural residents are integrated with succeedingly higher levels of community for various purposes
Relations within community	Primarily gemeinschaft	Gemeinschaft at lower levels of community integration but gesellschaft at higher levels
Family size	Large	Small
Family composition	Residential units nuclear but extended family ties important	Nuclear with declining importance of wider kinship ties
Authority structure of family	Patriarchal	Equalitarian
Internal relations and family solidarity	Based on "affective identification" among members	Based on "rational interdependence" among members
Family functions	Satisfied all needs of members including economic, social, educational, and recreational	Most functions taken over by other agencies; primarily satisfies emotional needs of members
Role of Religion	Pervasive; affecting almost all aspects of rural life	Increasingly differentiated from other areas of rural life
Role of the church	Highly important; often provided a structural focus for the rural community	Diminishing importance

TABLE 7.1 (cont'd)

Characteristics	Traditional	Contemporary
Number of churches	Numerous; "overchurching" was a typical feature	Fewer; diminishing rate of membership and attendance, and high cost of maintenance threatening its survival
Role of rural clergyman	Highly important, not only in religious but also in other spheres	Declining importance due to increasing secularization, and his inability to provide a dynamic leadership
Importance of and attitude toward formal education	Low and negative because of the "open" nature of farming as an occupation	High and positive because of the need of technical skills in farming and nonfarm occupations
Proportion of people with formal schooling	Low	High
Material resources for formal schooling	Poor; inadequately equipped and staffed one-room schools	Highly improved; well-equipped and better staffed consolidated schools
Financial self-sufficiency and autonomy of local governments	High	Low because of shrinking tax base and increasing financial dependence on senior governments

The traditional rural society in Canada, which existed even as late as the 1930s and early 1940s prior to the Second World War, was characterized by features somewhat resembling a peasant society. The majority of Canada's people lived in rural areas, and most of the rural people were farmers. Although there was some outmigration from rural to urban areas in search of a better life and income, its magnitude was not all that great. In demographic composition, the population was characterized by features which are typical of a rural society: high sex and dependency ratios, low rates of divorce and formal educational attainment, and a highly homogeneous ethnic composition.

Farming, which was the main occupation of the majority of the rural population, was primarily a subsistence, not a commercial, activity. Every effort was made to produce on the farm all that the family needed. Family farms were small in size, and were almost entirely dependent on unpaid family labor to produce a variety of crops with very little use of chemicals and fertilizers, and machinery. Although the aim of farmers was at securing for the needs of the family rather than to make a profit, some surplus was grown to sell directly to consumers for the needed cash to buy livestock or machinery. The majority of farmers were owner operators owning almost all the resources on the farm and did not borrow money unless it was absolutely necessary either for family purposes or to invest on the farm. The practice of the "virtues of temperance and parsimony" allowed the farmers to maintain an independence which they greatly cherished. Rural living and farming were highly valued and were followed almost like a religious calling. A farm youth started as a family laborer and worked his way up to being a farm owner, and stayed in that occupation for the rest of his life.

The traditional rural community was a community of farmers and farm families who spent most of their lives within their neighborhoods or "rangs," and occasionally visited the nearest village or town for shopping, trading, visiting, or for recreational purposes. The territory of economic and social activities of the rural residents was thus defined primarily by the boundaries of the town-country community to which they belonged.

The family in a traditional rural society was typically a farm family. It was often large in size, and sometimes included a married son and his children who worked on the farm under his father until he inherited the farm after his father's retirement or death. The authority structure of the farm family was patriarchal with father acting as the leader and decision-maker. The family was the most important of all groups to which a ruralite belonged because it satisfied almost all of his or her needs including economic, social, educational, religious, and recreational.

Religion had an all pervasive influence on traditional and rural life. It reinforced the values relating to agriculture and a rural way of life in moral and religious terms. The church was often the structural focus of a rural community with the rural clergyman as its leader both for spiritual and secular purposes. Indeed, the spiritual and secular spheres of life were often indistinguishable in the traditional rural society.

Although formal education was not highly valued in the traditional rural society because of the "openness" of agriculture as an occupation, almost all rural communities had a one-room school of its own run by the members of the community. The rural school, however, was inadequately equipped to offer instruction to students belonging to different grades with the help of a teacher who was often unprepared to take on such a horrendous task. The rural school suffered from a narrow curriculum and a low student retention rate. But besides being a place for educating the children of the community, the one-room school was also a community center where the residents of the community met for a variety of purposes. Indeed, the one-room school often constituted the structural nucleus of a neighborhood or a small community.

The local government in rural Canada enjoyed a high degree of autonomy because of its relative isolation and financial solvency. Since the base of tax was property rather than individuals and there were relatively few services which were to be offered to tax payers, the local government did not have to depend greatly on the senior governments for financial handouts. This enabled the local government in the traditional rural society to look after its own affairs without much external interference.

The traditional rural life of Canada was thus characterized by features typical of a rural society. Its relational structure was gemeinschaft with a preponderance of primary groups; it was primarily sacred in orientation with not a very great use of technical and rational knowledge in productive and other activities; and the solidarity of the communities and families were based on affective identification among members.

THE CONTEMPORARY RURAL SOCIETY

The contemporary rural society of Canada, however, appears to be far more urbanized when compared with the traditional (Table 7.1). The rural society in

Canada has become increasingly urbanized in demographic, economic, and social terms.

Only a small part of the Canadian population lives in rural areas today, a result of the increased urbanization and industrialization of the country. A minuscule proportion of the total population, and a small minority of the rural population, are farmers. The rural population in Canada is thus predominantly a nonfarm population. The "urbanization" of the rural population is also evident in the demographic composition. It increasingly resembles the urban population in sex and dependency ratios, rate of divorce, level of educational attainment, and ethnic composition.

Farming in rural areas is becoming an activity of a small group of commercially oriented entrepreneurs who operate very large-sized, fully or partly owned, incorporated farms with the help of modern machinery and most scientific techniques developed by agricultural researchers. These commercial farmers specialize in a few products to sell them for a profit in a market which is national and international in scope. They produce a lion's share of the country's total agricultural product and are increasingly expanding their operations by renting from or buying out small subsistence cultivators to remain competitive. Farming as an economic activity is no longer centered on the farm alone. It is rapidly becoming a part of the growing agrifood business which includes not only farmers but also a large variety of people off the farm who participate directly or indirectly in the production, processing, and marketing of agricultural products. The competitive pressure of the market has forced farmers, especially those with small and medium operations, to join rural cooperatives and farmers' organizations for a better bargaining position.

The technological transformation and commercialization of agriculture along with contact with the urban way of life, brought about by the improvement in transportation and communication, have profoundly affected the rural structure and institutions. The boundaries of the functional community of economic activities have expanded territorially with centralization of trading and services in larger centers although small communities continue to be the foci of affective relationships and symbolic identity. The contemporary rural family increasingly resembles its urban counterpart in size, composition, internal relationships, and functions. The rural-urban disparity in educational opportunity and attainment has narrowed drastically with school consolidation and changes in the attitudes of rural people toward formal education. The rural local government faces the same problems as the urban in maintaining its administrative and political autonomy while depending at the same time on provincial and federal governments for financial aid to provide modern public facilities and services to its tax payers.

The contemporary rural life in Canada thus has attained many features which are typical of an urban society. Its relational structure is increasingly gesellschaft with the rise in the number of secondary groups; it is increasingly becoming secular in orientation with the greater use of technical and instrumental principles in productive and other activities; and its solidarity is increasingly based on rational interdependence among groups which produce specialized goods and services.

SOURCES OF RURAL SOCIAL CHANGE

Although it is futile to look for a "prime mover" as a causal force of rural social change, for "causation in social change is multiple and reciprocal, not particularistic, simple and direct" (Ryan, 1969: 44), Redfield and Singer (1973: 346) identified technological development and wider contact with the outside world as two causal forces which move a society from the peasant (or rural) end toward the urban end of the rural-urban continuum. The transformation of rural Canada appears to have been primarily caused by the technological development in agriculture and increased contact of the rural people with the world outside, especially the urban world.

As has been noted, the technological development and mechanization have raised the productivity of agriculture and substituted capital for labor. The increased capitalization in agriculture has contributed to the movement of people from farming to nonfarm occupations, and from rural to urban areas. The decline of farming as the principal occupation of the majority of rural people has altered the very foundation on which the distinctive features of the rural society were based. The social and normative coherence of the traditional rural life centered around agriculture is in decline. The contemporary rural society is in the process of finding a new equilibrium based on reduced occupational importance of agriculture and involvement of its people predominantly in nonfarm activities. This is evident in the rise of individualism and weakening of traditional norms; increased division of labor and occupational specialization and consequent modification of social and economic organizations; the collective movements in the forms of cooperatives and farmers' organizations to protect the position and interests of farmers as a minority occupation group; and the "development of a more heterogeneous sociocultural matrix within which the individual's personality is formed, and within which he must work out for himself some reasonably comfortable pattern of accommodation" (Smith and Zopf, 1970: 497).

The loss of is relative isolation and increased contact of the members of the rural society with the world outside are also important contributing factors to the rural social change. Indeed, with the revolutionary improvement in the system of communication and transportation there has been a radical transformation "in the patterns of interaction within the community, as well as in the nature and frequency of contacts between communities" (Smith and Zopf, 1970: 495). Rural people travel back and forth between their communities and urban centers in working, trading, shopping, and visiting. The better roads, automobile, telephone, telegraph, and the electronic mass media make frequent contact of rural residents with townspeople and city dwellers possible. These have contributed to a rapid diffusion of urban traits and values to rural areas and a reduction of the rural-urban sociocultural differences. These urban traits and values manifest themselves not only in the spheres of work and economic activities but also in the size, composition, and relational norms within the rural family, segmentation of religion and secularization of rural norms and beliefs, and in the change of the educational system both in its material and normative dimensions.

FUTURE OF THE RURAL SOCIETY

Although the rural society in Canada continues to be small in population size and low in population density, it has moved far from a typical rural society in economic and sociocultural terms. The occupational shift from agriculture and diffusion of urban traits have given rise to "an urbanized population in dispersed settlements" (Coughenour and Busch, 1978: 212) of the countryside.

Does it mean that rural Canada is in the process of extinction? Does the trend point to an inevitable merger of the rural and urban "into one indistinguishable 'mass society'" (Ford, 1978: 41)? Several demographic and sociocultural trends indicate that such a situation is unlikely to occur. As has been noted, although the rural population has declined as a proportion of the total population, it has remained stable at around five million people since 1931. It is the farm population which has declined both in relative and absolute terms. More significantly, the rural population showed a proportional increase in 1981, for the first time in Canada's history. Much of this increase appears to be the result of reverse migration from metropolitan areas to outlying rural districts. The growth and diversification of rural economy and consequent broadening of occupational opportunities made it possible for those who prefer rural environment to stay in rural areas. Even though the rural population fell behind the urban population again in proportional increase in 1991, the decline was moderate and occurred only in the western provinces. Even the rate of decline of the farming population significantly slowed down in 1991.

Although the rural-urban sociocultural differences have narrowed considerably, it is remarkable that significant differences continue to persist in almost every sphere of social and cultural life. The different environments of rural and urban Canada continue to produce different behaviors and attitudes even after the dissipation of social forces that originally were thought to account for them. The rural-urban differences, while clearly diminishing, still remain crucial: "Urban life and rural life are still urban and rural despite their respective changes" (Sim, 1988: 59).

References

Abell, Helen C.

1961 "The Present Day Agricultural Ladder," *Economic Annalist*, vol. 31: 15-25.

1968 "The Social Consequences of Modernization of Agriculture," in Marc Adelard Tremblay and Walter J. Anderson (eds.), *Rural Canada in Transition*, Ottawa: Agricultural Economics Research Council, 178-227.

1975 "Adaptation of the Rural Family to Change," in Parvez Wakil (ed.), *Marriage, Family and Society: Canadian Perspectives*, Toronto: Butterworth, 367-78.

Abramson, J.A.

1968 *Rural to Urban Adjustment*, Saskatoon: Canadian Center for Community Studies, University of Saskatchewan.

Ackerman, Joseph and Marshall Harris (eds.)

1946 *Family Farm Policy*, Chicago: University of Chicago Press.

Alexander, Frank D.

1952 "Locality-Group Classification," *Rural Sociology*, vol. 16: 232-40.

Anderson, Allan

1977 *Remembering the Farm*, Toronto: Macmillan of Canada.

Anderson, Isabel

1966 *Internal Migration in Canada*, Ottawa: Economic Council of Canada.

Anderawewa, A.B.
1969 "Tenure Patterns and Commercialization of Canadian Agriculture," *Canadian Journal of Agricultural Economics*, vol. 17: 110-21.

Andrews, John H.M.
1954 *The Effect of Secondary School Consolidation upon Achievement in Fundamental and Unit Cost*, M.A. Thesis, Vancouver: University of British Columbia.

Appleton, Peter L.
1973 *The Canadian Agriculture and Food System*, Ottawa: Agricultural Economics Research Council of Canada.

Arensberg, Conrad M. and Arthur H. Niehoff
1964 *Introducing Social Change*, Chicago: Aldine Publishing.

Auer, L.
1975 *Canadian Agricultural Productivity*, Ottawa: Economic Council of Canada.

Bari, Alain and George A. Mori
1991 "Leaving the Fold: Declining Church Attendance," *Canadian Social Trends*, Autumn, Ottawa: Statistics Canada, 21-4.

Barrett, Stanley R.
1994 *Paradise: Class, Commuters, and Ethnicity in Rural Ontario*, Toronto: University of Toronto Press.

Barlett, Peggy F.
1986 "Part-time Farming: Saving the Farm or Saving the Lifestyle?," *Rural Sociology*, vol. 51: 289-313.

Basillais, Benoit
1999 "The Agri-Food Industry — It's Everywhere!" *Canadian Agriculture At a Glance*, Ottawa: Statistics Canada, Catalogue 96-325-XPB, 45-52.

Beal, George M., Everett M. Rogers, and Joe M. Bohlen
1957 "Validity of the Concept of Stages in the Adoption Process," *Rural Sociology*, vol. 22: 166-8.

Becker, Howard
1950 *Through Values to Social Interpretation*, Durham: Duke University Press.

Bell, Colin and Howard Newby
1972 *Community Studies: An Introduction to the Sociology of Local Community*, New York: Praeger Publishers.

Bender, Thomas
1975 *Toward an Urban Vision: Ideas and Institutions in Nineteenth Century America*, Lexington: University Press of Kentucky.

Bertrand, Alvin L. (ed.)
1958 *Rural Sociology: An Analysis of Contemporary Rural Life*, New York: MacGraw-Hill.

Bibby, Reginald W.
1993 *Unknown Gods: The Ongoing Story of Religion in Canada*, Toronto: Stoddart Publishing Co.

1995 *The Bibby Report: Social Trends Canadian Style*, Toronto: Stoddart Publishing Co.

Blackburn, Donald J. *et al.*
1975 *The Educational and Occupational Aspiration of Youth in Midnorthern Ontario*, Guelph: University of Guelph.

Bolaria, B. Singh
1992 "Farm Labour, Work Conditions and Health Risks," in David A. Hay and Gurcharan S. Basran (eds.), *Rural Sociology in Canada*, Toronto: Oxford University Press, 228-45.

Bollman, Ray D. and Pamela Smith
1988 "Integration of Canadian Farm and Off-farm Markets and the Off-farm Work of Women, Men, and Children," in G.S. Basran and D.A. Hay (eds.), *The Political Economy of Agriculture in Western Canada*, Toronto: Garamond Press, 185-202.

Bollman, Ray D. and Brian Biggs
1992 "Rural and Small Town Canada: An Overview," in Ray D. Bollman (ed.), *Rural and Small Town Canada*, Toronto: Thompson Educational Publishing, 3-44.

Bonner, Kieran
1997 *A Great Place to Raise Kids: Interpretation, Science and the Urban-Rural Debate*, Montreal and Kingston, Ont.: McGill-Queen's University Press.

Bose, Santi Priya
1964 "The Diffusion of a Farm Practice in Indian Villages," *Rural Sociology*, vol. 29: 53-66.

Boyd, Steve
1999 "All About Canada's Dairy Industry," *Canadian Agriculture At a Glance*, Ottawa: Statistics Canada, Catalogue 96-325-XPB, 71-9.

Brewster, John M.
1958 "Technological Advance and the Future of the Family Farm," *Journal of Farm Economics*, December: 1606-14.

Brickman, George L.
1980 "Reflection on Farm Income in the 1970s," *Canadian Journal of Agricultural Economics*, vol. 28: 56-66.

Brinkman, G.
1984 *An Analysis of Sources of Multifactor Productivity Growth in Canadian Agriculture, 1961 to 1980, with Projections to 2000*, Ottawa: Development Policy Directorate, Agriculture Canada.

Bristow-Callahan, Kim
1999 "Take Another Look at the Agriculture and the Census," *Canadian Agriculture At a Glance*, Ottawa: Statistics Canada, Catalogue 96-3256-XPB, 23-8.

Brown, Alan
1968 *Changing School Districts in Canada*, Toronto: Ontario Institute for Studies in Education.

Burnet, Jean
1951 *Next Year Country: A Study of Rural Social Organization in Alberta*, Toronto: University of Toronto Press.

Canadian Almanac and Directory
1999 "Municipal Directory," *Canadian Almanac and Directory 2000*, Section H, Toronto: Micromedia, 4.1-4.61.

Canadian Federation of Mayors and Municipalities
1976 *Puppets on a Shoestring: The Effects on Municipal Government of Canada's System of Public Finance*, Ottawa: Canadian Federation of Mayors and Municipalities.

The Canadian Institute of Public Opinions
1978 Gallup Polls, May 11
1979 Gallup Polls, July 28
1980 Gallup Polls, July 4
1983 Gallup Polls, July 11
1997 Gallup Polls, May 6
1998 Gallup Polls, July 12

Cape, E.
1984 "The Distaff Side of Retiring to the Country," presented at the Annual Meeting of Canadian Sociology and Anthropology Association, Guelph, Ontario.

Chamberlain, S.
1976 *Study of the Housing and Health Care Needs of Elderly Persons of North Frontenac*, Ottawa: School of Social Work, Carleton University.

Charles, Enid and Sylvia Anthony
1943 "The Community and the Family in Prince Edward Island," Rural Sociology, vol. 8: 37-51.

Cochrane, Jeane
1981 *The One Room School in Canada*, Toronto: Fitzhenry and Whiteside.

Connidis, I.
1989 *Family Ties and Aging*, Toronto: Butterworths.

Constandse, A.K.
1968 "Discussion" (on R.E. Pahl's, "Rural-Urban Continuum"), in R.E. Pahl (ed.), *Readings in Urban Sociology*, Oxford: Pergamon Press, 301-5.

Cooley, Charles
1909 *Social Organization*, New York: Scribner.

Cooperative College of Canada
1982 *Patterns and Trends in Canadian Cooperative Movement*, Saskatoon: Cooperative College of Canada.

Coughner, Milton and Lawrence Busch
1978 "Alternative Futures for Rural America: The Cloudy Crystal Ball," in Thomas R. Ford (ed.), *Rural U.S.A., Persistence and Change*, Ames: Iowa University Press, 211-46.

Crawford, Kenneth G.
1958 *Canadian Municipal Government*, Toronto: University of Toronto Press.

Culver, D. and R. Seecharan
1986 "Factors that Influence the Adoption of Soil Conservation Technologies," *Canadian Farm Economics*, vol. 20: 9-13.

Dahms, Fred
1996 "The Evolving Spatial Organization of Small Settlements in the Countryside — An Ontario Example," in Satadal Dasgupta (ed.), *The Community in Canada: Rural and Urban*, Lanham, Maryland: University Press of America, 200-11.

Dasgupta, Satadal
1982 "The Island in Transition: A Statistical Overview," in V. Smitheram, D. Milne, and S. Dasgupta (eds.), *The Garden Transformed: Prince Edward Island, 1945-1980*, Charlottetown: Ragweed Press, 243-68.

1984 "Modernization and Rural Community Organization: Changing Community Structure on Prince Edward Island," *International Review of Modern Sociology*, vol. 14: 18-32.

1985 *Caste, Kinship, and Community: Social System of a Bengal Caste*, London: Sangam Books.

1987 *Diffusion of Agricultural Innovations in Village India*, New Delhi: Wiley Eastern.

1996 *The Community in Canada: Rural and Urban*, Lanham, Maryland: University Press of America.

Dasgupta, Satadal and G.R. Madan
1978 *Community and Agriculture in Two Indian Villages*, Calcutta: Editions Indian.

Dasgupta, Satadal and P. Nagarajan
1977 *Factors in Educational Aspirations: A Study of High School Seniors in Prince Edward Island*, Charlottetown, Canada: University of Prince Edward Island, Department of Sociology and Anthropology.

Davis, A.R. and R. Goldberg
1959 *A Concept of Agribusiness*, Boston: Harvard University Press.

Davis, John H.
1956 "From Agriculture to Agribusiness," *Harvard Business Review*, vol. 34: 187-95.

Davis, Kingsley
1965 "The Urbanization of the Human Population," *Scientific American* (volume on cities), New York: Alfred A. Knopf, Inc.: 1-8.

Daviult, R.
1980 *Selected Agricultural Statistics*, Ottawa: Agriculture Canada.

Dawson, D.A. and K.J. Dancey
1976 "Economies of Scale in the Ontario Public School Sectors," *The Alberta Journal of Educational Research*, vol. 20: 182-8.

Dawson, Helen J.
1960 "An Interest Group: The Canadian Federation of Agriculture," *Canadian Public Administration*, vol. 3: 134-49.

Deffontaines, Pierre
1964 "The *Rang* — Pattern of Rural Settlement in French Canada," in Marcel Rioux and Yves Martin (eds.), *French-Canadian Society*, vol. 1, Toronto: McClelland and Stewart.

Dewey, Richard
1960 "Rural-Urban Continuum, Real but Relatively Unimportant," *American Journal of Sociology*, vol. 66: 60-6.

Diamant, Peter
1997 "Restructuring Local Government: Where Do We Go From Here," in Richard C. Rounds (ed.), *Changing Rural Institutions: A Canadian Perspective*, Brandon, Man.: Rural Development Institute, 61-72.

Dunlop, G.J. *et al.*
1957 "The Influence of Time Spent in School Buses upon Achievement and Attendance," *Alberta Journal of Educational Research*, vol. 3: 177-9.

Durkheim, Emile
1893 *De la division du travail social* (Translated by George Simpson, *The Division of Labor in Society*, New York: Macmillan, 1933).

Duncan, Otis D. and Albert Reiss, Jr.
1956 *Social Characteristics of Urban and Rural Communities, 1950*, New York: John Wiley.

Edmonds, E. and F. Bessai
1978 *Small Rural Schools on Prince Edward Island*, Charlottetown: Square Deal Publications.

Elkin, Frederick
1964 *The Rural Family in Canada*, Ottawa: Canadian Conference on the Family.

Fairbairn, Brett
1997 "Principles of Organizational Restructuring in Rural Organizations: Co-Operatives," in Richard C. Rounds (ed.), *Changing Rural Institutions: A Canadian Perspective*, Brandon, Man.: Rural Development Institute, 105-25.

Ford, Thomas R.
1978 "Contemporary Rural America: Persistence and Change," in Thomas R. Ford (ed.), *Rural U.S.A., Persistence and Change*, Ames: Iowa State University Press, 3-18.

Fortin, Gerald
1968 "The Challenge of a New World," in Marc Adelard Tremblay and Walter J. Anderson (eds.), *Rural Canada in Transition*, Ottawa: Agricultural Economics Research Council, 341-92.

1971 "Socio-Cultural Changes in an Agricultural Parish," in Marcel Rioux and Yves Martin (eds.), *French Canadian Society*, vol. 1, Toronto: McClelland and Stewart, 86-105.

Foster, George M.
1962 *Traditional Cultures and the Impact of Technological Change*, New York: Harper.

1967 "Introduction: What is a Peasant?" in Jack J. Potter *et al.* (eds.), *Peasant Society*, Boston: Little Brown, 2-14.

Fougere, Vernon
1997 "Catholic Churches Need Miracle," *The Northern Star*, August, 2.

Furniss, I.F.
1964 "Productivity of Canadian Agriculture, 1935-1960," *Canadian Journal of Agricultural Economics*, vol. 12: 41-53.

1970 "Agricultural Productivity in Canada: Two Decades of Gain," *Canadian Farm Economics*, vol. 5: 16-26.

Gagne, Stephan and Les McCagg
1995 *Co-Operatives in Canada*, Ottawa: Co-Operatives Secretariat, Government of Canada.

Galpin, C.J.
1915 *The Social Anatomy of an Agricultural Community*, Madison: Wisconsin Agricultural Experiment Station.

Gans, Herbert J.
1962 *The Urban Villagers*, New York: The Free Press.

Garigue, Philippe
1956 "French Canadian Kinship and Urban Life," *American Anthropologist*, vol. 58: 1096-1112.

1971 "Change and Continuity in Rural French Canada," in Marcel Rioux and Yves Martin (eds.), *French Canadian Society*, vol. 1, Toronto: McClelland and Stewart, 123-36.

Gerin, Leon
1971 "The French Canadian Family: Its Strengths and Weaknesses," in Marcel Rioux and Yves Martin (eds.), *French Canadian Society*, vol. 1, Toronto: McClelland and Stewart, 32-56.

Glock, Charles Y., Benjamin B. Ringer, and Earl R. Babic
1967 *To Comfort and to Challenge*, Berkeley: University of California Press.

Griezic, Foster J.K.
1975 "Introduction" in Louis A. Wood, *A History of Farmers' Movements in Canada*, Toronto: University of Toronto Press.

Gubbles, Peter M.
1966 *Adoption and Rejection of Innovation by Dairy Men in the Lower Fraser Valley*, M.S.A. Thesis, Vancouver: University of British Columbia.

Gunn, T.C.
1978 "Part-time Farming in Nova Scotia," *Canadian Farm Economics*, vol. 13: 25-41.

Haga, W.J. and C.L. Folse
1971 "Trade Pattern and Community Identity," *Rural Sociology*, vol. 36: 42-51.

Halpern, Joel M.
1967 *The Changing Village Community*, New York: Prentice-Hall.

Harper, S.
1987 "The Kinship Network of the Rural Aged: A Comparison of the Indigenous Elderly and the Retired Immigrant," *Aging and Society*, 7: 303-27.

Harrison, Rick and Sylvain Cloutier
1995 *People in Canadian Agriculture*, Ottawa: Statistics Canada, Catalogue 21-523E.

Hart, J.F.
1975 *The Look of the Land*, Englewood Cliffs, N.J.: Prentice-Hall.

Hauser, Philip M.
1967 "Observations on the Urban-Folk and Urban-Rural Dichotomies as forms of Western Ethnocentrism," in Philip M. Hauser and Leo F. Schnore (eds.), *The Study of Urbanization*, New York: John Wiley, 503-14.

Herberle, R.
1938 "The Causes of Rural-Urban Migration: A Survey of German Theories," *American Journal of Sociology*, vol. 43: 932-50.

Hightower, Jim
1978 "Agribusiness and Agrigovernment: Power, Business and Poverty," in Richard D. Rodefeld *et al*. (eds.), *Change in Rural America*, St. Louis: C.V. Mosby, 278-82.

Hill, C.E.
1976 "Canadian Cooperatives: Their Growth and Finance," *Canadian Farm Economics*, vol. 11: 8-9.

Hillery, George A.
1955 "Definitions of Community: Areas of Agreement," *Rural Sociology*, vol. 20: 111-23.

Hodge, G.
1984 *Shelter and Services for the Small Town Elderly: The Role of Assisted Housing*, Ottawa: Canada Mortgage and Housing Corporation.

Hodge, G.D. and M.A. Qadeer
1983 *Towns and Villages in Canada*, Toronto: Butterworth.

Hofstee, E.W.
1960 "Rural Social Organization," *Sociologia Ruralis*, vol. 1: 105-17.

Hughes, Charles C. *et al.*
1960 *People of Cove and Woodlot: Communities from the Viewpoint of Social Psychiatry*, New York: Basic Books.

Janssen, William
1982 "Agriculture in Transition," in V. Smitheram, D. Milne, and S. Dasgupta (eds.), *The Garden Transformed: Prince Edward Island, 1945-1980*, Charlottetown: Ragweed Press, 115-29.

Jean, Bruno
1997 "Issues in Institutional Restructuring in the Rural Sector," in Richard C. Rounds (ed.), *Changing Rural Institutions: A Canadian Perspective*, Brandon, Man.: Rural Development Institute, 5-19.

Jones, Gwyn
1967 "The Adoption and Diffusion of Agricultural Practices," *World Agricultural Economics and Rural Sociology Abstracts*, vol. 9: 1-34.

1973 *Rural Life*, London: Longman.

Kalbach, Warren E. and Wayne W. McVey
1979 *The Demographic Bases of Canadian Society*, Toronto: McGraw-Hill Ryerson.

Katz, Elihu, Martin L. Levin, and Herbert Hamilton
1963 "Tradition of Research on the Diffusion of Innovation," *American Sociological Review*, vol. 28: 237-52.

Keating, Norah

1987 "The work of Farm Women," *First Reading*, 5: 5-7.

1992 "Older Rural Canadians," in David A. Hay and Gurcharan S. Basran (eds.), *Rural Sociology in Canada*, Toronto: Oxford University Press, 134-50.

Keane, Michael J.

1978 "Economics and the Size of Rural Schools," *Canadian Journal of Agricultural Economics*, vol. 26: 47-52.

Kohl, Seena and John W. Bennett

1974 "Succession to Family Enterprises and the Migration of Young People in an Agricultural Community," in S. Crysdale and C. Beattie (eds), *Sociology Canada: Readings*, Toronto: Butterworth, 203-25.

Kralt, John

1976 *The Urban and Rural Composition of Canada's Population*, 1971 Census of Canada Profile Studies, Bulletin 5.1-2, Ottawa: Information Canada.

Kroeber, Alfred L.

1948 *Anthropology*, New York: Harcourt, Brace & Co.

Krout, J.A.

1988 "The Elderly in Rural Environments," *Journal of Rural Studies*, vol. 4: 103-4.

Larsen, Vernon W.

1964a *The Minister and the Church*, Saskatoon: Center for Community Studies, University of Saskatchewan.

1964b *A Venture in Church Cooperation*, Saskatoon: Center for Community Studies, University of Saskatchewan.

Laskin, Richard

1968 "Nonagricultural, Semiagricultural, and Agricultural Service Centers," in Marc Adelard Tremblay and Walter J. Anderson (eds.), *Rural Canada in Transition*, Ottawa: Agricultural Economics Research Council, 114-77.

Lee, T.

1957 "On the Relations Between the School Journey and Emotional Adjustment in Rural Infant Children," *British Journal of Educational Psychology*, vol. 40: 101-14.

Levitt, Keri
1960 *Population Movement in the Atlantic Provinces*, Halifax: Atlantic Provinces Economic Council.

Lewis, G.J.
1979 *Rural Communities*, London: David and Charles.

Lewis, Oscar
1967 "Further Observations on the Folk-Urban Continuum and Urbanization with Special Reference to Mexico City," in Philip M. Hauser and Leo F. Schnore (eds.), *The Study of Urbanization*, New York: John Wiley, 491-503.

Liebersohn, Harry
1988 *Fate and Utopia in German Sociology, 1870-1923*, Cambridge, Mass.: Massachusetts Institute of Technology Press.

Lionberger, Herbert F.
1960 *Adoption of New Ideas Practices*, Ames: Iowa State University Press.

Lipset, Seymore
1955 "Social Mobility and Urbanization," *Rural Sociology*, vol. 20: 220-8.

Lonmo, Charlene
1999 "Online On the Farm," *Canadian Agriculture At a Glance*, Ottawa: Statistics Canada, Catalogue 96-325-XPB, 239-44.

Lupri, Eugene
1968 "Discussion" (on R.E. Pahl's "The Rural-Urban Continuum"), in R.E. Pahl (ed.), *Readings in Urban Sociology*, Oxford: Pergamon Press, 298-300.

Lyons, R.M.A. and M. Kraut
1992 "The Family Farm in the Next Decade: The Positive Role of Part-time Farming," *Canadian Journal of Agricultural Economics*, vol. 40: 591-604.

MacDonald, Allan F.
1960 *A Sociological Study of the Parish of Portage*, M.A. Thesis, Washington, D.C.: Catholic University of America.

MacDougal, John
1973 *Rural Life in Canada*, Toronto: University of Toronto Press (first published, 1913).

MacGregor, Pat
1999 "Experts are Key to Canada's Hog Industry," *Canadian Agriculture At a Glance*, Ottawa: Statistics Canada, Catalogue 96-325-XPB, 107-11.

MacPherson, Ian
1982 "The Heritage," *Patterns and Trends in Canadian Cooperative Movement*, Saskatoon: Cooperative College of Canada, 6-38.

Mann, W.E.
1948 *The Rural Church in Canada*, Toronto: The Christian Social Council of Canada.

Martin Mathews, A.
1988 "Aging in Rural Canada," in E. Rathbone-McCuan and B. Havens (eds.), *North American Elders: United States and Canadian Perspectives*, Westport, CT: Greenwood, 143-60.

McCagg, Les
1999 *Co-Operatives in Canada*, Ottawa: Co-Operatives Secretariat, Government of Canada.

McKinney, John C. and Charles P. Loomis
1963 "The Application of Gemeinschaft and Gesellschaft as Related to Their Typologies," in Charles P. Loomis (ed.), *Community and Society*, New York: Harper and Row (Torchbook edition).

McVey, Wayne W., Jr. and Warren Kalbach
1995 *Canadian Population*, Toronto: Nelson Canada.

Meyer, Bruce
1992 "Population, Income and Migration Characteristics for Urban/Rural Areas and Farm/Nonfarm Families in Saskatchewan," in Ray D. Bollman (ed.), *Rural and Small Town Canada*, Toronto: Thompson Educational Publishing, 299-318.

Millard, Frank W.
1965 *An Analysis of the Adoption of Innovations by Okanagan Orchardists*, M.S.A. Thesis, Vancouver: University of British Columbia.

Miner, Horace

1939 *St. Denis, A French-Canadian Parish*, Chicago: University of Chicago Press.

1971 "Changes in Rural French Canadian Culture," in Marcel Rioux and Yves Martin (eds.), *French Canadian Society*, vol. 1, Toronto: McClelland and Stewart, 634-75.

Mol, Hans

1976 "Major Correlates of Churchgoing in Canada," in Stewart Crysdale and Les Wheatcraft (eds.), *Religion in Canadian Society*, Toronto: MacMillan of Canada, 241-54.

Mori, George A.

1987 "Religious Affiliation in Canada," *Canadian Social Trends*, Autumn, Ottawa: Statistics Canada, 12-16.

Morton, M., S. Polowin, C. Murphy, and A. McDonald

1984 *Survey of Needs of Seniors in the Township of Rideau*, Rideau, Ont.: Township of Rideau Senior Citizens' Service Centre.

Negrave, Patti

1999 "Try the Alternative Way," *Canadian Agriculture At a Glance*, Ottawa: Statistics Canada, Catalogue 96-325-XPB, 93-7.

Norland, Joseph A.

1976 *The Age-Sex Structure of Canada's Population*, 1971 Census Canada Profile Studies, Bulletin 5.1-3, Ottawa: Information Canada.

Overbreak, Johannes

1980 *Population and Canadian Society*, Toronto: Butterworth.

Pahl, R.E.

1968 "The Rural-Urban Continuum," in R.E. Pahl (ed.), *Readings in Urban Sociology*, Oxford: Pergamon Press, 263-97.

Peabody, George

1992 *School Days: One Room Schools of Maritime Canada*, Fredericton, NB: Gooselane Editions.

Plourde, Robert

1999 "The North American Beef Market: Competition Keeps it Lean," *Canadian Agriculture At a Glance*, Ottawa: Statistics Canada, Catalogue 96-325-XPB, 99-105.

Porter, John

1961 *The Vertical Mosaic*, Toronto: University of Toronto Press.

Randhawa, B.S.

1992 "Rural Education: Opportunities and Issues," in David A. Hay and Gurcharan S. Basran (eds.), *Rural Sociology in Canada*, Toronto: Oxford University Press, 101-12.

Rainey, Kenneth D. and Karen G. Rainey

1978 "Rural Government and Local Public Services," in Thomas R. Ford (ed.), *Rural U.S.A.: Persistence and Change*, Ames: Iowa State University, 126-44.

Redfield, Robert

1941 *The Folk Culture of Yucatan*, Chicago: University of Chicago Press.

1943 "Rural Sociology and the Folk Society," *Rural Sociology*, vol. 8: 68-71.

1947 "The Folk-Urban Continuum," *American Journal of Sociology*, vol. 52: 292-308.

1953 *Primitive World and Its Transformations*, Ithaca, New York: Cornell University Press.

1956 *Peasant Society and Culture*, Chicago: University of Chicago Press.

Redfield, Robert and Milton Singer

1971 "City and Countryside: The Cultural Independence," in Teodor Shanin (ed.), *Peasant Societies*, Middlesex: Penguin Books (originally published in *Economic Development and Social Change*, vol. 3, 1954: 53-73), 337-65.

Reimer, Bill

1986 "Women as Farm Labor," *Rural Sociology*, vol. 51: 143-55.

1997 "Informal Social Networks and Voluntary Associations in Non-Metropolitan Canada," in Richard C. Rounds (ed.), *Changing Rural Institutions: A Canadian Perspective*, Brandon, Man.: Canadian Rural Development Institute, 87-104.

Reiss, Albert J., Jr.
1955 "An Analysis of Urban Phenomena," in Robert M. Fisher (ed.), *The Metropolis in Modern Life*, Garden City, N.Y.: Doubleday and Co., 42-55.

Rioux, Marcel
1971 "Kinship Recognition and Urbanization in French Canada," in Marcel Rioux and Yves Martin (eds.), *French Canadian Society*, vol. 1, Toronto: McClelland and Stewart, 372-85.

Rodefeld, Richard D.
1978 "The Causes of Change in Farm Technology, Size and Organizational Structure," in Richard D. Rodefeld *et al.* (eds.), *Change in Rural America*, St. Louis: C.V. Mosby Co., 217-37.

Rogers, Everett M.
1962 *Diffusion of Innovations*, New York: The Free Press.

Rogers, Everett M. and Rabel J. Burdge
1972 *Social Change in Rural Societies*, New York: Appleton-Century-Crofts.

Rogers, Everett M. and F. Floyd Shoemaker
1971 *Communication of Innovations*, New York: The Free Press.

Rogers, Everett M. *et al.*
1988 *Social Change in Rural Societies: An Introduction to Rural Sociology*, Third Edition, Englewood Cliffs, N.J.: Prentice-Hall.

Rogers, Walter B. and George E. Buckmire
1967 *The Rural Church, The Rural Family*, Agricultural Economics, Report 7, Edmonton: University of Alberta.

Ross, Romaien K.
1962 *Local Government in Ontario*, Toronto: Canada Law Book Co.

Rowat, Donald C.
1975 *Your Local Government*, Toronto: MacMillan of Canada.

Royal Commission on Agriculture and Rural Life
1956 *Rural Education*, Report 6, Regina: Queen's Printer, Government of Saskatchewan.

1956 *The Home and Family in Rural Saskatchewan*, Report 10, Regina: Queen's Printer, Government of Saskatchewan.

Ryan, Bryce F.
1969 *Social and Cultural Change*, New York: Ronald Press.

Scheidt, R.J.
1984 "A Taxonomy of Well-Being for Small-Town Elderly," *The Gerontologist*, 24: 84-90.

Selles, R. and N. Keating
1989 "La Transmission des Fermes par les âges d'origine Hollandaise," in R. Santerre and D. Meintel (eds.), *Veiller au Quebec, en Afrique et Ailleurs*, Laval: University of Laval Press.

Shaver, F.M.
1990 "Women, Work and Transformations in Agricultural Production," *Canadian Review of Sociology and Anthropology*, vol. 27: 341-56.

Shute, D.M.
1975 "National and Regional Productivity of Canadian Agriculture, 1961-1974," *Canadian Farm Economics*, vol. 10: 1-6.

Siegel, David
1994 "The ABCs of Local Government: An Overview," in Dale Richmond and David Siegel (eds.), *Agencies, Boards and Commissions in Canadian Local Government*, Monographs on Canadian Public Administration of Canada — No. 15, Toronto: The Institute of Public Administration of Canada, 1-20.

Siemens, Leonard B.
1965 *The Influence of Selected Family Factors on the Educational and Occupational Aspiration Levels of High School Boys and Girls*, Winnipeg: University of Manitoba.

Sim, Alex R.
1988 *Land and Community: Crisis in Canada's Countryside*, Guelph, Ont.: University of Guelph.

1990 *The Plight of the Rural Church*, McGeachy Papers, Volume II, Toronto: The United Church Publishing House.

Sinclair, Sol
1958 "Discussion: Technological Advance and the Future of the Family Farm," *Journal of Farm Economics*, vol. 20: 1601-21.

Slocum, Walter L.
1962 *Agricultural Sociology*, New York: Harper and Brothers.

Smith, Pamela
1987 "What Lies Within and Behind the Statistics? Trying to Measure Women's Contribution to Canadian Agriculture," *Growing Strong: Women in Agriculture*, Ottawa: The Canadian Advisory Council on the Status of Women, 123-208.

1992 "Beyond 'Add Women and Stir' in Canadian Rural Society," in David A. Hay and Gurcharan S. Basran (eds.), *Rural Sociology in Canada*, Toronto: Oxford University Press, 155-70.

Smith, T. Lynn and Paul E. Zopf, Jr.
1970 *Principles of Inductive Rural Sociology*, Philadelphia: F.A. Davis Company.

Smitheram, Verner
1975 "School Consolidation on P.E.I.: A Mixed Blessing," *Abegweit Review*, vol. 1: 50-9.

1982 "Development and Debate over School Consolidation," in V. Smitheram, D. Milne, and S. Dasgupta (eds.), *The Garden Transformed: Prince Edward Island, 1945-80*, Charlottetown: Ragweed Press, 177-202.

Sorokin, Pitrim and Carle C. Zimmerman
1929 *Principles of Rural and Urban Sociology*, New York: Holt, Rinehart, and Winston.

Statistics Canada
1979 *Canada's Families*, Catalogue 98-801E, Ottawa: Statistics Canada.

1989 *Canada Year Book 1990*, Ottawa: Statistics Canada.

1999 *Canadian Agriculture At a Glance*, Catalogue 96-325-XPB, Ottawa: Statistics Canada, Agriculture Division.

Stone, Leroy O.
1961 *Internal Migration*, Census Monograph, Ottawa: Queen's Printer.

Stone, L.O. and S. Fletcher
1980 *A Profile of Canada's Older Population*, Montreal: The Institute for Research on Public Policy.

Sullivan, J.M.
1982 *Cooperation in Canada*, Ottawa: Agriculture Canada.

1993 *Co-Operation in Canada*, Ottawa: Co-Operatives Secretariat, Government of Canada.

Task Force on Canadian Agriculture
1977 *Orientation of Canadian Agriculture: A Task Force Report*, vols. 1 to 3, Ottawa: Agriculture Canada.

Thurston, N.E. *et al.*
1982 "Health Status of the Rural Elderly: A Picture of Health," paper presented at the Annual Meeting of the Canadian Association of Gerontology, Winnipeg, Manitoba.

Tindal, C.R. and S. Nobes Tindal
1979 *Local Government in Canada*, Toronto: McGraw-Hill Ryerson.

Trant, M.J. and G.L. Birkman
1979 "A Classification of Limited Resource Farmers," *Canadian Farm Economics*, vol. 14: 21-56.

Tonnies, Ferdinand
1887 *Gemeinschaft und Gesellschaft*, Auflage, Leipzig (translated by Charles P. Loomis, *Community and Society*, East Lansing: Michigan University Press, 1957).

Tung, F.L. and G. Strain
1987 "Research, Technology and Productivity Change in Canadian Agriculture," *Canadian Farm Economics*, vol. 21: 37-43.

Veeman, Terry S.
1995 "Agriculture and Resource Economics: Challenges for the 21st Century," *Canadian Journal of Agricultural Economics*, vol. 34: 519-28.

Verdon, Michel
1980 "The Quebec Stem Family Revisited," in K. Ishwaran (ed.), *Canadian Families: Ethnic Variations*, Toronto: McGraw-Hill Ryerson, 105-24.

Verner, C. and F.W. Mildred
1966 *Adult Education and Adoption of Innovations by Orchardists in the Okanagan Valley of British Columbia*, Vancouver: University of British Columbia.

Verner, Coolie and Peter M. Gubbles
1967 *The Adoption or Rejection of Innovations by Dairy Farm Operators in the Lower Fraser Valley*, Ottawa: Agricultural Economics Council

Vivian, J.B.
1982 *Home Support Services Survey Project*, St. John's: Newfoundland Department of Social Services.

Wall, Ellen
1987 "Offshore Farm Workers in Rural Ontario Communities," *The Rural Sociologist*, vol. 7: 453-8.

Wall, G., E. Vaughan, and G. Driver
1985 *Cropping, Tillage and Land Management Practices in Southwestern Ontario 1984*, Toronto: Ontario Institute of Pedology.

Weissman, Ernest
1967 "The Great Migration to the City," in Hans Singer *et al.* (eds.), *International Development*, New York: Oceania Publications, 40-52.

Wenger, C.
1982 "Aging in Rural Communities: Family Contacts and Community Integration," *Aging and Society*, 2: 211-29.

Westhues, Kenneth and Peter Sinclair
1974 *Village in Crisis*, Toronto: Holt, Rinehart, and Winston of Canada.

Whyte, Donald R.
1968 "Rural Canada in Transition," in Marc Adelard Tremblay and Walton J. Anderson (eds.), *Rural Canada in Transition*. Ottawa: Agricultural Economics Research Council, 1-113.

Windley, P.G. and R. Scheidt
1988 "Rural Small Towns: An Environmental Context for Aging," *Journal of Rural Studies*, 4: 151-8.

Wirth, Louis
1938 "Urbanism as a Way of Life," *American Journal of Sociology*, vol. 44: 1-24.

Wilkinson, Kenneth P.
1978 "Rural Community Change," in Thomas R. Ford (ed.), *Rural U.S.A.: Persistence and Change*, Ames: Iowa University Press, 115-25.

Wolf, Eric R.
1966 *Peasants*, Englewood Cliffs, N.J.: Prentice-Hall.

Wood, Louis A.
1975 *A History of Farmers' Movements in Canada*, Toronto: University of Toronto Press.

Zimmerman, Carl C. and Gary Moneo
1970 "The Total Community of the Canadian Wheat Prairies," *Canadian Agricultural Economics*, vol. 18: 6-28.

Index

CANADIAN STUDIES

1. Clifford J. Jansen, **Italians in a Multicultural Canada**
2. Dieter Hoehne, **Legal Aid in Canada**
3 Satadal Dasgupta, **Rural Canada: Rural Sociological Perspectives**
4. Mark Mullins, **Religious Minorities in Canada: A Sociological Study of the Japanese Experience**
5. Joan Vance, **An Analysis of the Costs and Benefits of Public Lotteries: The Canadian Experience**
6. Peter A. Russell, **Attitudes To Social Structure and Mobility in Upper Canada 1815-1840: "Here We Are Laird Ourselves"**
7. John S. Brierley and Daniel Todd, **Prairie Small-Town Survival: The Challenge of Agro-Manitoba**
8. Scott Bennett, (ed.), **Technology and Work in Canada**
9. Yvan Gagnon, **The Theory and Practice of Public Policy-Making in Canada: Metapolicymaking**
10. John Cooke, **The Influence of Painting on Five Canadian Writers: Alice Munro, Hugh Hood, Timothy Findlay, Margaret Atwood, Michael Ondaatje**
11. Kenneth W. Meadwell, ***L'avalée des avalés, L'hiver de force* et *Les enfantômes* de Réjean Ducharme: Une fiction mot à mot et sa littérarité**
12. John Benson, **Entrepreneurism in Canada: A History of "Penny Capitalism"**
13. Lorna Berman and Irina Sobkowska-Ashcroft, **Portrayal of Old Age in Twentieth Century Canadian Novels**
14. Michael Pomedli, **William Kurelek's Huronia Mission Paintings**
15. James Horne Morrison, **The Diary of James Horne Morrison, (1894-1895), A Scots Missionary in Canada**, edited and introduced by Boyd Stanley Schlenther
16. Karin M. Egloff, **Le thème du regard dans l'oeuvre d'André Langevin, écrivain québécois: la flèche dans l'oeil de Narcisse**
17. John M. Naish, **The Interwoven Lives of George Vancouver, Archibald Menzies, Joseph Widbey and Peter Puget**
18. James M. Dean, **What Type of Financial Institutions Would be Needed to Incorporate the Northwest Territories into the Canadian Federation?: Fiscal Policies for the Future**

19. Robert I. McLaren, **The Saskatchewan Practice of Public Administration in Historical Perspective**

20. Daniel A. Metraux, **The Soka Gakkai Buddhist Movement in Quebec**

21. Eva Darias-Beautell, **Contemporary Theories and Canadian Fiction: Shifting Sands**

22. John C. Pierce, Nicholas P. Lovrich, Brent S. Steel, Mary Ann E. Steger, John R. Tennert, **Political Culture and Public Policy in Canada and The United States: Only A Border Apart?**

23. Stéphanie Nutting, **Le tragique dans le théâtre québécois et canadien-français, 1950-1989**

24. Alexander Kizuk, **A Reassessment of Early Twentieth-Century Canadian Poetry in English**

25. Martin Thornton, **The Domestic and International Dimensions of the Resettlement of Polish Ex-Servicemen in Canada, 1943-1948**

26. Michael Dennis McGowan, **The Canonical Status of Catholic Health Care Facilities in the Province of New Brunswick in Light of Recent Provincial Government Legislation**

27. Rosalee Stilwell, **The *Ethos* of Voice in the Journal of James Rainstorpe Morris From the Sable Island Humane Station, 1801-1802**

28. Susan E. Billingham, **Language and the Sacred in Canadian Poet bpNichol's *The Martyrology***

29. Kenneth J. Munro, **A Biography of François-Xavier-Anselme Trudel, Quebec's Foremost Political Maverick in the Nineteenth Century**